Springer Laboratory

Manuals in Polymer Science

Series editors

More information about this series at http://www.springer.com/series/3721

Brunero Cappella

Mechanical Properties of Polymers Measured through AFM Force-Distance Curves

Springer

Brunero Cappella
BAM
Federal Institute of Materials Research
and Testing
Berlin, Germany

ISSN 0945-6074 ISSN 2196-1174 (electronic)
Springer Laboratory
ISBN 978-3-319-29457-5 ISBN 978-3-319-29459-9 (eBook)
DOI 10.1007/978-3-319-29459-9

Library of Congress Control Number: 2016934847

Printed on acid-free paper

This Springer imprint is published by Springer Nature
The registered company is Springer International Publishing AG Switzerland

Contents

Symbols

a	Contact radius	m
$\overline{A} = \frac{a}{\sqrt[3]{\pi W R^2 / E_{\text{tot}}}}$	Dimensionless contact radius	
a_0	Contact radius at zero load	m
A	Area	m^2
A_H	Hamaker constant	J
a_T	Shift factors (WLF or Arrhenius equation)	
c	Concentration or number density	number/m^3 mol/m^3 mol/L
C_1, C_2	Coefficients of the Williams–Landel–Ferry equation	°C (C_2)
D	Deformation, distance	m
$\overline{D} = \frac{D}{\sqrt[3]{\pi^2 W^2 R / E_{\text{tot}}^2}}$	Dimensionless deformation	
D_m	Molecular diameter	m
$D_n \cong \sqrt[3]{R W^2 / E_{\text{tot}}^2}$	Height of the neck formed by negative loads	m
D_p	Deformation at zero load	m
E	Young's modulus	Pa
E_c (E_t)	Young's modulus of the cantilever (of the tip)	Pa
E'	Storage modulus	Pa
E''	Loss modulus	Pa
E_0	Instantaneous modulus	Pa
E_∞	Equilibrium modulus	Pa
E_a	Activation energy	J
E_D	Dissipated energy	J
E_{tot}	Reduced elastic modulus	Pa
F	Force	N
$\overline{F} = F/\pi W R$	Dimensionless force	
F_{adh}	Adhesion force	N
G	Shear modulus	Pa

(continued)

G, G_{mix}	Gibb's free energy, Gibb's free energy of mixing	J
h	Indentation	m
H, H_{mix}	Enthalpy, enthalpy of mixing	J
$I_c = w_c t_c^3/12$	Moment of inertia of the cantilever	m^4
$J = 1/E$	Compliance	Pa^{-1}
K	Bulk modulus	Pa
k_c	Elastic constant of the cantilever	N/m
$k_{\text{eff}} = k_c k_s/(k_c + k_s)$	Effective elastic constant of the system	N/m
k_s	Elastic constant of the sample	N/m
l	Length of a monomer	m
l_p	Plastic zone size	m
L	Length	m
L_c	Length of the cantilever	m
m	Mass	g
$m_M = a/\chi$	Ratio of the contact radius to the radius of the annular region where the adhesion is accounted for	
M_n	Mean weight	kg/mol, kDa
M_w	Molecular weight	kg/mol, kDa
n	Refractive index	
n_m	Number of monomers in a polymer chain	
N	Polymerisation degree	
P	Pressure	Pa, N/m^2
Q	Quality factor of the cantilever	
R	(tip) radius	m
R_{eff}	Effective radius	m
R_F	Flory radius	m
R_g	Gyration radius of a polymer	m
S	Stiffness	N/m
$S_{\text{eff}} = k_s/(k_c + k_s)$	Effective stiffness of the system	N/m
S, S_{mix}	Entropy, mixing entropy	J/K
t	Time	s
t_c	Thickness of the cantilever	m
t_c	Contact time	s
t_f	Thickness of a polymer film or layer	m
T	Temperature	K
T_g	Glass transition temperature	K
T_m	Melting temperature	K
$\tan\delta$	Loss factor	
U	Internal energy	J
V	Volume	m^3
v_f, V_f	Excess free volume	m^3
$v_F = \mathrm{d}F/\mathrm{d}t$	Loading rate	N/s

(continued)

V_m	Molecular volume	m^3
v_p	Displacement velocity of the contact line	m/s
W	Adhesion work per unit area	N/m
w_c	Width of the cantilever	m
z_0	Typical atomic dimension	m
Z	Piezo displacement	m
Greek letters		
α_c	Opening angle of a V-shaped cantilever	rad
$\alpha_V = \mathrm{d}V/\mathrm{d}T$	Thermal expansion coefficient	m^3/K
$\gamma_c\ (\gamma_s)$	Damping coefficient of the cantilever (sample)	$N \cdot s/m$
Γ	Grafting density	number/m^2
δ	Cantilever deflection	m
Δ_S	Displacement of the reflected spot	m
ε	Dielectric permittivity	
ε_b, ε_s, ε_t	Bulk, shear and tensile strain	
ζ	Valence	
θ	Semi-aperture of a conical punch	rad
$\lambda = \frac{2.06}{z_0}\sqrt{\frac{RW^2}{\pi E_{\mathrm{tot}}^2}}$	Maugis parameter	
$\lambda_D = \sqrt{\frac{\varepsilon\varepsilon_0 k_B T}{e^2 \sum c_i^0 \zeta_i^2}}$	Debye length	m
Λ	Wavelength	m
μ	Chemical potential	J/mol
μ_M	Müller's parameter	
$\nu\ (\nu_t)$	Poisson's ratio (of the tip)	
ν_e	Absorption frequency	Hz
ρ	Mass density	kg/m^3
σ	Normal stress	N/m^2
σ_A	Parameter of Attard and Parker	
$\sigma_S\ (\sigma_T)$	Charge density of the sample (of the tip)	C/m^2
σ_y	Yield strength	N/m^2
τ	Shear stress	N/m^2
τ	Retardation time	s
τ_r	Relaxation time	s
φ	Phase	rad
φ_P	Pashley's parameter	
Π	Volume fraction	
$\Phi(t)$	Creep compliance	
χ	Radius of the annular region where the adhesion is accounted for	m
χ_{12}	Flory interaction parameter	
ω	Angular frequency	Hz
ω_0	Angular resonance frequency of the cantilever	Hz
Ω	Sensitivity	V/m

(continued)

Fundamental constants		
h	Planck constant	6.626×10^{-34} Js
k_B	Boltzmann constant	1.38×10^{-23} J/K
N_A	Avogadro constant	6.022×10^{23} mol^{-1}
R_g	Gas constant	8.31 J/(mol K)

Acronyms

ABCPA	4,4′-azobis(4-cyanopentanoic acid)
AFM	Atomic force microscope
DLVO	Derjaguin–Landau–Verwey–Overbeek
DMA	Dynamic mechanical analysis
DMSO	Dimethylsulfoxide
DMT	Derjaguin–Müller–Toporov
DSC	Differential scanning calorimetry
GPS	(3-Glycidoxypropyl) trimethoxysilane
HDPE	High-density polyethylene
HWHM	Half-width at half-maximum
JKR	Johnson–Kendall–Roberts
LDPE	Low-density polyethylene
LVDT	Linear variable displacement transformers
μTA	Microthermal analysis
PAA	Polyacrylic acid
PB	Polybutadiene
PnBMA	Poly(*n*-butyl methacrylate)
PtBuA	Poly(*tert*-butyl acrylate)
PC	Polycarbonate
PDMS	Poly(dimethyl siloxane)
(PEN)	Poly(ethylene naphthalate)
PET	Poly(ethylene terephthalate)
PFM	Pulsed force mode
PI	Polyisoprene
PMA	Poly(methyl acrylate)
PMMA	Poly(methyl methacrylate)
PS	Polystyrene
PSF	Poly(styrene-*co*-pentafluorostyrene)
PMA	Polymethacrylate
PP	Polypropylene
PU	Polyurethane
PVA	Polyvinyl alcohol
PVC	Polyvinyl chloride

PVDF	Polyvinylidene fluoride
SEM	Scanning electron microscope
SFM	Scanning force microscope
SLS	Standard linear solid
SNOM	Scanning near-field optical microscope
SPM	Scanning probe microscopes
STM	Scanning tunnelling microscope
UHMWPE	Ultrahigh molecular weight polyethylene
WLF	Williams–Landel–Ferry

Part I

Principles: Theory and Practice

Physical Principles of Force–Distance Curves by Atomic Force Microscopy

1

Abstract

The atomic force microscope (AFM) is increasingly employed not only to acquire topography images of samples but also to measure force–distance curves. Such curves, beyond playing a major role in the theoretical study of surface interactions, are meanwhile a fundamental tool in surface science, nanotechnology, biology and many other fields of research.

Force–distance curves find their application in the study of numerous material properties, such as mechanical properties, surface charge densities, adhesion and Hamaker constants.

One of the most important applications of AFM force–distance curves is the study of mechanical properties of polymers. Compared to other instruments, the AFM has in this case two major advantages. First of all, elastic moduli of samples can be measured with high resolution from some GPa down to some MPa, which is the range of the elastic moduli of common polymers. Second, force–distance curves can be acquired in an array over the sample. This is a fundamental tool for the characterization of the lateral variation of sample properties and hence for the study of confined polymers and polymer blends.

The first part of this book is divided in two chapters dealing with the theoretical and practical aspects of force–distance curves. Theoretical aspects, handled in this chapter, are focused on mechanical properties of polymers.

1.1 Atomic Force Microscope

The atomic force microscope (AFM), also known as scanning force microscope (SFM), invented by Binnig, Quate and Gerber in 1986 [1], is not only a microscope permitting to image sample surfaces with atomic resolution but also a tool permitting to investigate several properties of the samples, in particular mechanical properties.

B. Cappella, *Mechanical Properties of Polymers Measured through AFM Force-Distance Curves*, Springer Laboratory, DOI 10.1007/978-3-319-29459-9_1

The AFM belongs together with its predecessors, the scanning tunnelling microscope (STM) [2] and the scanning near-field optical microscope (SNOM) [3, 4], to the group of the scanning probe microscopes (SPM). The common operative principle of these microscopes is the investigation of a sample through a probe scanned in the near field of a certain physical effect. The STM exploits the quantum–mechanical tunnelling effect and the SNOM operates in the optical near field. The probe of the AFM scans the near field of the surface forces acting between probe and sample.

The core of the AFM is a cantilever with a micro-fabricated tip, which deflects when interacting with the sample [5]. AFM cantilevers are usually made out of silicon or silicon nitride. The most common cantilevers can be divided into two categories depending on their shape: rectangular and V-shaped. The cantilever back face, i.e. the face that is not in contact with the sample, is usually coated with a metallic thin layer, often gold, in order to enhance the reflectivity. This is necessary in liquids, where the reflectivity of silicon and silicon nitride is much reduced. Further details about AFM cantilevers are given in Sect. 2.2.

In order to image the sample topography and to acquire force–distance curves, the sample must be scanned in *X*-, *Y*- and *Z*-direction by a piezoelectric actuator, which must be able to perform with high-precision minimal displacements of the order of 1 Å up to displacements of several tens of microns. Unfortunately, the dependence of the displacement of the piezo on the applied voltage is hysteretic and affected by creep, i.e. a delay effect depending on temperature. Because of creep, a little fraction of the displacement is performed with logarithmic time dependence. The most used method to eliminate the nonlinearities of piezoelectric actuators is to measure independently their displacements with inductive sensors [6].

For the measurement of the cantilever deflection, several methods have been developed. The most commonly used method is the optical lever technique [7]. It consists in focusing a laser beam on the backside of the cantilever and detecting the reflected beam by means of a position sensor, usually a photodiode consisting of four quadrants. With this method both cantilever deflections and torsions can be detected. The optical lever technique is illustrated in detail in Sect. 2.1.

For the imaging of sample surfaces, the AFM can be operated in several modes, depending on the way the tip and/or the sample are scanned during the measurement and on the signals acquired. Since the main subject of this book is the study of mechanical properties of polymer samples through force–distance curves, the two most used modes for topography imaging, contact mode and Tapping Mode, are introduced here very briefly. A complete review of imaging with AFM can be found in [8]. Both theoretical and practical aspects of imaging of polymer samples are reviewed in [9].

In contact mode the tip is always in contact with the sample. During the scan, the cantilever deflection is kept constant through a feedback system. Since tip and sample are in contact, the constant value of the deflection, i.e. the setpoint, is positive. In other words, the force acting onto the cantilever is repulsive. In order to keep the deflection constant, the tip or the sample is moved vertically, i.e. perpendicularly to the sample surface. The topography of the sample is reconstructed from the changes in the piezo elongation.

Since in contact mode tip and sample are always in contact, also during lateral movements, the cantilever undergoes not only deflections but also torsions. Torsions are caused by a nonzero slope of the sample topography or by friction. Hence, in contact mode, the AFM can be used also to determine friction coefficients or for mapping differences in the friction of the sample [10].

Because of such lateral forces between tip and sample, it may happen that some structures on the sample surface are removed (dragging) or the sample is damaged. Therefore, contact mode is usually not suitable for imaging of soft materials like polymers.

In Tapping Mode the surface of the sample is scanned while the cantilever oscillates at a frequency near its resonance frequency ω_0. The excitation for the oscillation is provided by an extra piezoelectric transducer (dither piezo). Let the free oscillation amplitude, i.e. the oscillation amplitude when the tip is far away from the sample, be some tens of nanometres. If the tip–sample distance is reduced, the tip touches the sample at the bottom of each oscillation cycle and is both in the attractive and in the repulsive regime during each oscillation period. As a consequence, the oscillation amplitude is reduced. Analogously to contact mode, the oscillation amplitude is kept constant, and the topography of the sample is reconstructed from the piezo elongation, which is necessary to keep the oscillation amplitude equal to the setpoint.

Tapping Mode is employed most of all with soft and compliant samples, because the tip does not exert any lateral forces and the sample is not damaged.

1.2 Force–Distance Curves

A force–distance curve [11, 12] is a graph of the force experienced by the AFM cantilever versus the distance between the AFM tip and the sample surface. In order to record a force–distance curve, tip and sample are approached to each other till contact and then separated. Depending on the particular microscope, this is achieved by moving the tip or the sample in *Z*-direction, i.e. perpendicular to the sample surface, by means of a piezoelectric transducer. While varying the tip–sample distance, two signals are acquired: the deflection of the cantilever and the displacement of the piezoelectric transducer. Hence, prior to the conversion of the signals, a force–distance curve should be called more properly a deflection–displacement curve. The conversion of the deflection into a force and of the displacement into the tip–sample distance is explained below.

According to the two directions of movement, a force–distance curve encompasses two parts, the approach and retraction part. Both parts of force–distance curves can be roughly divided into three regions: the zero line, the discontinuity and the contact region.

The zero line is recorded when the tip–sample distance is so large that no measurable force is acting between tip and sample.

The discontinuity, called jump-to-contact in the approach curve and jump-off-contact in the retraction curve, occurs at the tip–sample distance, at which the gradient of the total attractive force exceeds the elastic constant of the cantilever (approach), or the elastic constant of the cantilever exceeds the gradient of the total adhesion force (retraction), so that the tip snaps abruptly onto the sample surface or detaches from it.

The contact line is the part of the curve along which tip and sample are in contact and the tip is pushed against the sample and possibly indents it.

Under the realistic assumption that the deflection of the cantilever is much smaller than its length, we can assume that the cantilever is elastic and behaves like an ideal spring. In this case the applied force F is given by Hooke's law:

$$F = -k_c \delta, \tag{1.1}$$

where k_c is the spring constant of the cantilever and δ the cantilever deflection. The deflection and the force are considered positive when the cantilever bends away from the sample (repulsive force) and negative when the cantilever bends towards the sample (attractive force).

If the deflection and the spring constant of the cantilever are known (see Sect. 2.1 for the measurement of the deflection and Sect. 2.2.1 for the calibration of the spring constant), Eq. (1.1) can be used to convert the deflection into a force.

For the calculation of the tip–sample distance D, the following equation is used:

$$D = Z - \delta. \tag{1.2}$$

Here, Z is the distance between the sample surface and the rest position of the cantilever. If $\delta > Z$ the distance turns out to be negative, i.e. the tip indents the sample and the deformation into the sample surface is positive.

In order to understand how to convert the displacement of the piezoelectric transducer, Z, into the tip–sample distance D, it is necessary to consider a force–displacement curve as the result of two contributions: the tip–sample interaction $F(D)$ and the elastic force of the cantilever $F = -k_c\delta$ (Eq. 1.1). The interplay of these two contributions can be understood through a graphical construction of a force–displacement curve. Such a graphical construction yields not only a procedure for the mentioned conversion but also an explanation for the occurrence of the discontinuities.

In Fig. 1.1 the curve $F(D)$ is the tip–sample interaction force. Since the fundamental results of this graphical construction do not depend on the chosen force law, a Lennard–Jones force law is used in the form

$$F(D) = -\frac{A}{D^2} + \frac{B}{D^6}. \tag{1.3}$$

The straight lines 1–4 with slope equal to the spring constant of the cantilever, k_c, represent the opposite of the elastic force of the cantilever, $F_{\mathrm{el}} = D_0 + k_c D$.

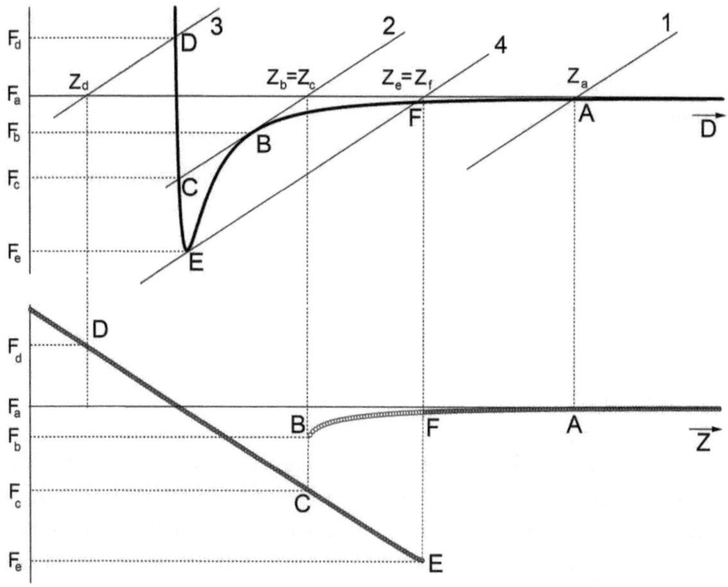

Fig. 1.1 Graphical construction of a force–displacement curve. In the *top panel*, the tip–sample interaction $F(D)$ is shown. The lines 1–4 with a slope equal to the spring constant of the cantilever, k_c, represent the opposite of the elastic force of the cantilever. At each distance the cantilever deflects down to the equilibrium position, i.e. the distance at which the elastic force counterbalances the tip–sample force. The force values F_a till F_f are the ordinates of the intersections A–F between $F(D)$ and the lines 1–4 (F_f, which is nearly zero, is not shown for clarity). In the graphical construction of the force–displacement curve, these force values are assigned to the distances Z between the sample and the rest position of the cantilever, i.e. the distances Z_a till Z_f given by the intersections between the lines 1–4 and the horizontal axis. The result is shown in the *bottom panel*. The approach curve, obtained shifting a line $F_{\text{el}}=D_0+k_cD$ from right to left, is plotted with *empty circles*; the retraction curve, obtained shifting the line F_{el} from left to right, is plotted with *grey circles*

The resulting force–displacement curve is shown in the bottom panel (empty circles are the approach curve, grey points the retraction curve). At each distance the cantilever deflects till it reaches the equilibrium position, i.e. the distance D at which the elastic force of the cantilever counterbalances the tip–sample interaction force. Hence, the approach curve is obtained by letting a straight line $F_{\text{el}}=D_0+k_cD$ run from right to left (i.e. from point A to point D); for the retraction curve, F_{el} runs from left to right (i.e. from point D back to point A over points E and F). The force values at equilibrium, F_a till F_f, are given by the intersections A–F between the lines 1–4 and the curve $F(D)$ (F_f, which is nearly zero, is not shown for clarity). During the acquisition of a force–displacement curve, the tip–sample distance D is unknown. The only measured distances are the distances Z between the rest positions of sample and cantilever. As a consequence, the equilibrium force values are not assigned to the respective distances D, but to the distances Z_a till Z_f, i.e. the intersections between the lines 1–4 and the zero-force axis, given by $Z = D + \delta = D - F/k_c$.

Some important issues can be understood intuitively through this graphical construction.

1. At the points B in the approach and E in the retraction curve, there is no stable equilibrium: an infinitesimal change in the distance lets the intersection between the line F_{el} and the curve $F(D)$ "jump" from the point B to the point C (or from the point E to the point F), which have the same abscissa in the bottom graph. This is the origin of the jump-to-contact and jump-off-contact. The discontinuities can be avoided only when the elastic constant of the cantilever, and hence the slope of the line F_{el}, is larger than the maximum of the first derivative of $-F(D)$.
2. If k_c is smaller than the maximum of the first derivative of $-F(D)$, i.e. in the presence of discontinuities, the force–distance curve shows a hysteresis: in the approach the curve goes from A to D over B and C, in the retraction over E and F. Moreover, during the approach, the interval of the function $F(D)$ between the points B and C is not sampled; during the retraction the not sampled interval is that between the points E and F. This hysteresis depends on k_c and can be avoided with a very stiff cantilever.
3. The zero distance of a force–displacement curve, and hence its origin, is the intersection between the zero line and the contact line.
4. If the repulsive part of the curve $F(D)$ is very steep, i.e. $-\partial F/\partial D \gg k_c$, the slope of the contact line is nearly $-k_c$ and the slope of the contact line of a deflection–displacement curve is nearly -1.
5. In order to convert the displacements measured during the acquisition of a force–displacement curve into the tip–sample distance, it is sufficient to perform the inverse graphical construction. In this case the result can be expressed analytically. Given a certain point I of the force–displacement curve with coordinates (Z_i, F_i), it is necessary to find the intercept with the axis $F = 0$ of the line with slope $-k_c$ passing through I, i.e. $F = -k_c(Z - Z_i) + F_i$. The intercept is hence given by $Z = F_i/k_c + Z_i$ and this is the distance D_i corresponding to the force F_i. Since during the approach the cantilever jumps from point B to point C and during the retraction curve from point E to point F, the portion of the curve $F(D)$ between point B and point E cannot be reconstructed.

Considering now the contact line of a force distance curve, a quantity playing a major role in the description of the mechanical properties of samples can be introduced, namely, the effective elastic constant k_{eff}.

Along the contact line, the tip–sample distance D equals the opposite of the deformation into the sample. We assume that the sample is elastic and can be modelled as an ideal spring, i.e. its stiffness $k_s = \partial F/\partial D$ is a constant. This can be the case only for small deformations. Thanks to the balance of forces during contact, Eq. (1.1) can be written as

$$k_c|\delta| = k_s|D|, \tag{1.4}$$

where k_s is the elastic constant of the sample.

Equations (1.2) and (1.4) yield

$$F = k_c\delta = \frac{k_c k_s}{k_c + k_s} Z = k_c S_{\text{eff}} Z = k_{\text{eff}} Z. \tag{1.5}$$

The effective spring constant k_{eff} can be calculated with a simple linear fit of the contact line and is a good indicator for the sample stiffness, even if it is obtained by rather rough assumptions.

Depending on the ratio of k_c and k_s, further approximations can be done. Equation (1.5) can be written as $\delta = Z$ for $k_s \gg k_c$ (i.e. $S_{\text{eff}} = 1$) and as $\delta = (k_s/k_c)Z$ for $k_s \ll k_c$.

The first approximation for not deformable samples is essential for the calculation of the sensitivity (see Sect. 2.1).

For a further examination of the contact line of force–distance curves, leading to a quantitative determination of mechanical quantities as Young's modulus, the sample stiffness cannot be considered as a constant and the sample deformation has to be analysed with an elastic continuum theory. Elastic continuum theories are reviewed in Sect. 1.4.

1.3 Elasticity and Storage Elastic Modulus

In this chapter some basic definitions of continuum mechanics [13], mainly the definition of elastic modulus, are given.

When a force F is applied on a surface of area A, the stress is defined as the force per unit area. In general, a stress engenders a deformation of the body. There are two basic kinds of deformations: shear deformations, i.e. a change in shape with no change in volume, and bulk compression or dilatation, i.e. a change in volume with no change in shape.

Shear deformations are due to coplanar forces, i.e. forces applied tangentially to the body surface. The corresponding stress is called the shear stress τ. Bulk deformations are due to normal forces. The corresponding stress is called the normal stress σ.

The strain is defined as the relative deformation of a body. In case of a change in length, for example, when a tensile or compressive stress is applied only to the ends of a rod and not to the sides, the tensile strain is given by

$$\varepsilon_t = \frac{\Delta L}{L_0}, \tag{1.6}$$

where ΔL is the change in length and L_0 is the initial length.

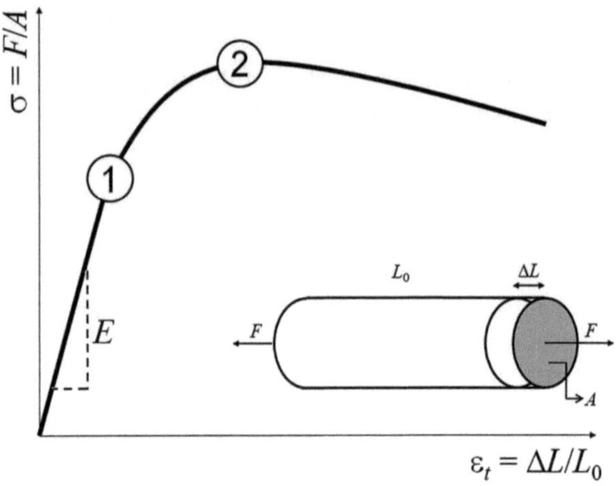

Fig. 1.2 Schematic stress–strain curve for a rod, shown in the *bottom part*. Points 1 and 2 are the elastic limit and the yield strength

When the stress is applied to all sides of a body and the whole volume is changed, the bulk strain is defined as

$$\varepsilon_b = \frac{\Delta V}{V_0}, \tag{1.7}$$

where ΔV is the change in volume and V_0 is the initial volume.

A graph of the stress as a function of deformation is called the stress–strain curve. A schematic example is shown in Fig. 1.2 for the deformation of a rod.

The first part of the curve, where stress and strain are proportional, is called the linear elastic region. In this region a modulus can be defined as the ratio of the stress to the strain. In case of a tensile strain, the modulus is called tensile modulus or Young's modulus $E = \sigma/\varepsilon_t$; in case of a bulk strain, it is the bulk modulus $K = \sigma/\varepsilon_b$; in the presence of a shear deformation, the modulus is called shear modulus $G = \tau/\varepsilon_s$, where ε_s is the shear strain.

Shear and bulk or length deformations are related, because a change in shape can affect also the volume and vice versa. The relation between E and G can be expressed in terms of a dimensionless parameter, Poisson's ratio ν. For example, when applying a stress to the ends of a rod, not only the length L changes but also the diameter of the cross section d. Poisson's ratio is defined as

$$\nu = -\frac{\Delta d/d_0}{\Delta L/L_0}. \tag{1.8}$$

The volume strain is related to Poisson's ratio through

$$\varepsilon_b = \frac{\Delta V}{V_0} = (1 - 2\nu)\frac{\Delta L}{L_0} = (1 - 2\nu)\varepsilon_t. \tag{1.9}$$

The definition of Poisson's ratio yields following relations for the moduli:

$$\nu = \frac{E}{2G} - 1 = \frac{1}{2} - \frac{E}{6K} = \frac{3K - 2G}{6K + 2G}. \tag{1.10}$$

After the linear elastic region, i.e. after point 1 in Fig. 1.2, there is a region where the deformation is still reversible, i.e. elastic, but no longer proportional to the stress; this part of the curve is called the nonlinear elastic region. The last region of the stress–strain curve is the plastic region. In this region the deformation is partially irreversible (see Sect. 1.9).

1.4 Indentation and Continuum Elastic Theories

As outlined in Sect. 1.2, the contact line of force–distance curves can be described approximately by assuming that the sample can be modelled as an ideal spring and its stiffness is constant and in particular does not depend on the load.

Yet, most samples cannot be described as an ideal spring; the stiffness of the indenter–sample system, k_s, is not a constant and the load is not proportional to the deformation.

The general definition of k_s is given by [14]

$$k_s = \frac{\partial F}{\partial D} = \frac{3}{2} a E_{\text{tot}}, \tag{1.11}$$

where a is the contact radius.

The reduced elastic modulus E_{tot} is given by

$$\frac{1}{E_{\text{tot}}} = \frac{3}{4}\left(\frac{1 - \nu^2}{E} + \frac{1 - \nu_t^2}{E_t}\right), \tag{1.12}$$

where E and ν (E_t and ν_t) are Young's modulus and Poisson's ratio of the sample (of the tip). Note that if the tip is much stiffer than the sample ($E_t \gg E$), which is often the case, Eq. (1.12) can be written as

$$k_s = 2a\left(\frac{E}{1 - \nu^2}\right). \tag{1.13}$$

A definition of the function $a(F)$, and hence of $k_s(F)$ and, via the relation $D = \int \frac{1}{k_s} dF$, also of the function $D(F)$, is given by elastic continuum theories.

Elastic continuum theories have two important limitations, as their name already says: they are valid only if the deformation of the sample is totally elastic, and they can predict the deformation only if the interacting bodies are considered as a continuum, i.e. only when the molecular structure and displacements of single molecules are ignored.

The three basic elastic continuum theories are those of Hertz [15], Johnson–Kendall–Roberts (JKR) [16] and Derjaguin–Müller–Toporov (DMT) [17–19].

The major difference between the three theories is given by the adhesion, which is neglected in Hertz theory and accounted for, but in different ways, in JKR and DMT theories.

1.4.1 Hertz Theory

Hertz theory, published in 1881, deals actually with the deformation of an elastic sphere pressed onto a rigid flat plane. The analogous, inverse problem of the deformation of an elastic flat plane indented by a sphere has been studied by Sneddon [20]. Sneddon's analysis can be applied to any solid of revolution.

Deformation and force are given in the absence of adhesion by

$$F = 2E_{\text{tot}}a\int_0^1 \frac{x^2 f'(x)}{\sqrt{1-x^2}}\text{d}x, \tag{1.14}$$

$$D = \int_0^1 \frac{f'(x)}{\sqrt{1-x^2}}\text{d}x, \tag{1.15}$$

in which $f(x)$ is a function describing the profile of the indenting punch.

For a sphere of radius R on a plane, $f(x) = R - \sqrt{R^2 - a^2x^2}$ and Eqs. (1.14) and (1.15) become

$$F = \frac{3}{4}E_{\text{tot}}\left[\frac{a^2+R^2}{2}\ln\left(\frac{R+a}{R-a}\right) - aR\right], \tag{1.16}$$

$$D = \frac{a}{2}\ln\left(\frac{R+a}{R-a}\right). \tag{1.17}$$

If deformations are much smaller than the radius of the sphere, we get

$$a^3 = \frac{RF}{E_{\text{tot}}}, \tag{1.18}$$

$$D = \frac{a^2}{R} = \left(\frac{F}{E_{\text{tot}}\sqrt{R}}\right)^{2/3}. \tag{1.19}$$

Equations (1.18) and (1.19) are valid also for the load and the deformation of two spheres of radii R_1 and R_2, provided we substitute to R the effective radius R_{eff}, given by

$$R_{\mathrm{eff}} = \frac{R_1 R_2}{R_1 + R_2}. \tag{1.20}$$

Given a punch having the form of a solid of revolution, Eq. (1.19) can be written as $F = \alpha D^n$. The coefficients are $\alpha = \frac{3}{2\pi} E_{\mathrm{tot}} \tan\theta$ and $n = 2$ for a cone (being θ the semi-aperture of the cone) and $\alpha = \frac{3}{2} R E_{\mathrm{tot}}$ and $n = 1$ for a flat-ended cylinder of radius R. For a paraboloid whose apex has a radius of curvature R, the valid equation is Eq. (1.19). Moreover, for a paraboloid, Eq. (1.19) is not limited to small deformations and is valid in general.

In Hertz and Sneddon approach, the adhesion between punch and sample is neglected. Hence, this theory can be applied only when the maximum adhesion between the AFM tip and the sample is much smaller than the maximum load. This is the case for several polymers with Young's moduli $E > 1$ GPa and large yielding force, i.e. the force at which plastic deformations start occurring. In such cases the polymer can be indented up to large forces without plastic deformations and the small adhesion force can be neglected, when compared with the high loads.

1.4.2 Derjaguin–Müller–Toporov and Johnson–Kendall–Roberts Theories

Both DMT and JKR theory consider the adhesion force acting between the punch and the sample. When the adhesion is not negligible, tensile stresses are added to compressive stresses, which are the only ones considered in Hertz theory. Hence, the adhesion force contributes to the load and alters the contact radius.

The DMT theory takes into account only the forces acting between the two bodies outside the contact region. These forces alone produce a finite area of contact, which increases by applying an additional external load. Hence, the contact radius at zero load, a_0, other than in Hertz theory, is not zero and is given by

$$a_0^{\mathrm{DMT}} = \sqrt[3]{\frac{2\pi R^2 W}{E_{\mathrm{tot}}}}, \tag{1.21}$$

in which W is the adhesion work per unit area.

If the external load is negative, the contact area decreases until it reaches zero. At this point, the adhesion force reaches its maximum value, given by

$$F_{\mathrm{adh}} = 2\pi R W. \tag{1.22}$$

The corresponding equations for the load and the deformation show that DMT theory can be derived from Hertz theory by adding a factor $2\pi R W$ to the load:

$$a^3 = \frac{R}{E_{\mathrm{tot}}}(F + 2\pi RW), \tag{1.23}$$

$$D = \frac{a^2}{R} = \left(\frac{F + 2\pi RW}{E_{\mathrm{tot}}\sqrt{R}}\right)^{2/3}. \tag{1.24}$$

JKR theory results from the opposite approach as that of DMT theory: long-range forces outside the contact area are neglected and only short-range forces inside the contact region are considered. For the contact radius at zero load, we get

$$a_0^{\mathrm{JKR}} = \sqrt[3]{\frac{6\pi R^2 W}{E_{\mathrm{tot}}}}. \tag{1.25}$$

The maximum adhesion force, the contact radius and the deformation are given by

$$F_{\mathrm{adh}} = \frac{3}{2}\pi RW, \tag{1.26}$$

$$a^3 = \frac{R}{E_{\mathrm{tot}}}\left(F + 3\pi RW + \sqrt{6\pi RWF + (3\pi RW)^2}\right), \tag{1.27}$$

$$D = \frac{a^2}{R} - \frac{2}{3}\sqrt{\frac{6\pi Wa}{E_{\mathrm{tot}}}}. \tag{1.28}$$

In JKR theory, during unloading, a neck is formed and contact is ruptured at a negative load $F = -F_{\mathrm{adh}} = -\frac{3}{2}\pi RW$, when the contact radius and the deformation assume the values

$$a = a_{\mathrm{min}} = \sqrt[3]{\frac{3\pi R^2 W}{2E_{\mathrm{tot}}}} = \frac{a_0}{\sqrt[3]{4}}, \tag{1.29}$$

$$D = D_{\mathrm{min}} = -\sqrt[3]{\frac{\pi^2 RW^2}{12E_{\mathrm{tot}}^2}}. \tag{1.30}$$

Hence, in JKR theory, negative deformations, i.e. deformations out of the sample surface and not into the sample surface, are possible. As a consequence, JKR theory predicts a hysteresis of the load–deformation curves. For negative values of the load, there are two possible values of the deformation and also two possible values of the contact radius for negative deformations. The actual deformation and the actual contact radius depend on the loading history.

JKR and DMT theories predict two fundamentally different deformation behaviours of the sample. The question, which theory is the "true" one, has been debated in the 1970s and 1980s, often very vehemently. The result of this controversy was that both theories are true, but apply to different systems.

The deciding parameter in the question, which theory must be applied to a certain system, is the height of the neck formed for negative loads, which is also the maximum negative deformation:

$$D_n \cong \sqrt[3]{\frac{RW^2}{E_{\mathrm{tot}}^2}}. \tag{1.31}$$

If the neck can be neglected, i.e. for small punch radii, low adhesion and stiff samples, DMT theory should be applied. In case of large punch radii, large adhesion and compliant samples, i.e. when the neck is higher, JKR theory is more suitable.

The height of the neck is proportional to a number of parameters which have been proposed along the years to "decide" between the two theories, such as the parameter σ_A of Attard and Parker [21], the parameter μ_M of Müller [18] and the parameter φ_P of Pashley [22].

The controversy on the two theories has been composed by Maugis through a novel elastic continuum theory.

1.4.3 Maugis Theory

Maugis [23, 24] has shown that DMT and JKR theories are approximations and limits of a unique theory, in which the adhesion is considered in an annular region of radius χ around the contact area.

In Maugis theory the deformation of the sample is described as a function of the parameter λ given by

$$\lambda = \frac{2.06}{z_0}\sqrt{\frac{RW^2}{\pi E_{\mathrm{tot}}^2}}, \tag{1.32}$$

where z_0 is a typical atomic dimension. Also this parameter is proportional to the height of the neck D_n.

In the Maugis theory, the deformation and the contact radius are given by a set of parametric equations:

$$\overline{D} = \overline{A}^2 - \frac{4}{3}\lambda\overline{A}\sqrt{m_M^2 - 1}, \tag{1.33}$$

$$\overline{F} = \overline{A}^3 - \lambda\overline{A}^2\left(\sqrt{m_M^2 - 1} + m_M^2\alpha\right), \tag{1.34}$$

where the dimensionless deformation, contact radius and force are given by

$$\overline{D} = \frac{D}{\sqrt[3]{\pi^2 W^2 R / E_{tot}^2}}, \ \overline{A} = \frac{a}{\sqrt[3]{\pi W R^2 / E_{tot}}}, \ \overline{F} = \frac{F}{\pi W R}, \tag{1.35}$$

and

$$\lambda \overline{A}^2 \left[\sqrt{m_M^2 - 1} + \alpha (m_M^2 - 2) \right] + \frac{4\lambda^2 \overline{A}}{3} \left(1 - m_M + \alpha \sqrt{m_M^2 - 1} \right) = 1. \tag{1.36}$$

The parameter $m_M = a/\chi$ is the ratio of the contact radius to the radius of the annular region where the adhesion is accounted for and $\alpha = \arctan\sqrt{m_M^2 - 1}$.

With these notations the equations for $a(F)$ and $D(F)$ in DMT theory become

$$\overline{A} = \sqrt[3]{(\overline{F} + 2)}, \tag{1.37}$$

$$\overline{D} = \sqrt[3]{(\overline{F} + 2)^2} = \overline{A}^2. \tag{1.38}$$

In JKR theory

$$\overline{A} = \sqrt[3]{\overline{F} + 3 + \sqrt{6\overline{F} + 9}}, \tag{1.39}$$

$$\overline{D} = \sqrt[3]{\overline{F} + 3 + \sqrt{6\overline{F} + 9}^2} - \frac{2}{3}\sqrt{6\sqrt[3]{\overline{F} + 3 + \sqrt{6\overline{F} + 9}}} = \overline{A}^2 - \frac{2}{3}\sqrt{6\overline{A}}. \tag{1.40}$$

Figure 1.3 shows the contact radius as a function of the deformation (A), the deformation as a function of the force (B) and the contact radius as a function of the force (C) in Hertz (red), DMT (blue), JKR (black) and in Maugis theory for different values of λ ($\lambda = 0.01$, 0.1, 0.5, 1 and 2, dashed lines). The curves have been calculated with $R = 20$ nm and $E_{tot} = 3$ GPa. For JKR and DMT curves, $W = 0.25$ N/m has been used.

It is clear that Maugis theory reduces to DMT theory for $\lambda \to 0$ and to JKR theory for $\lambda \to \infty$. For $\lambda > 1$ there are two values of the deformation for negative values of the load (Fig. 1.3B) and also two values of the contact radius for negative deformations (Fig. 1.3A) or for negative loads (Fig. 1.3C), and the curves are hysteretic.

At this point the hysteresis predicted by JKR theory can be understood better. Figure 1.4 shows again the force as a function of deformation for JKR theory. It is the same curve as in Fig. 1.3B, but with exchanged axes. The straight lines L_1–L_3 represent the elastic force of a cantilever with spring constant k_c. As in the graphical construction outlined in Sect. 1.2, the force values at equilibrium are given by the intersections 1–4 between the lines L_1–L_3 and the JKR curve. In a force–distance curve, the equilibrium force values are assigned to the distances Z, i.e. to the intersections between the lines L_1–L_3 and the zero-force axis. As for the jump-to-contact and the jump-off-contact in Sect. 1.2, during the approach, at point

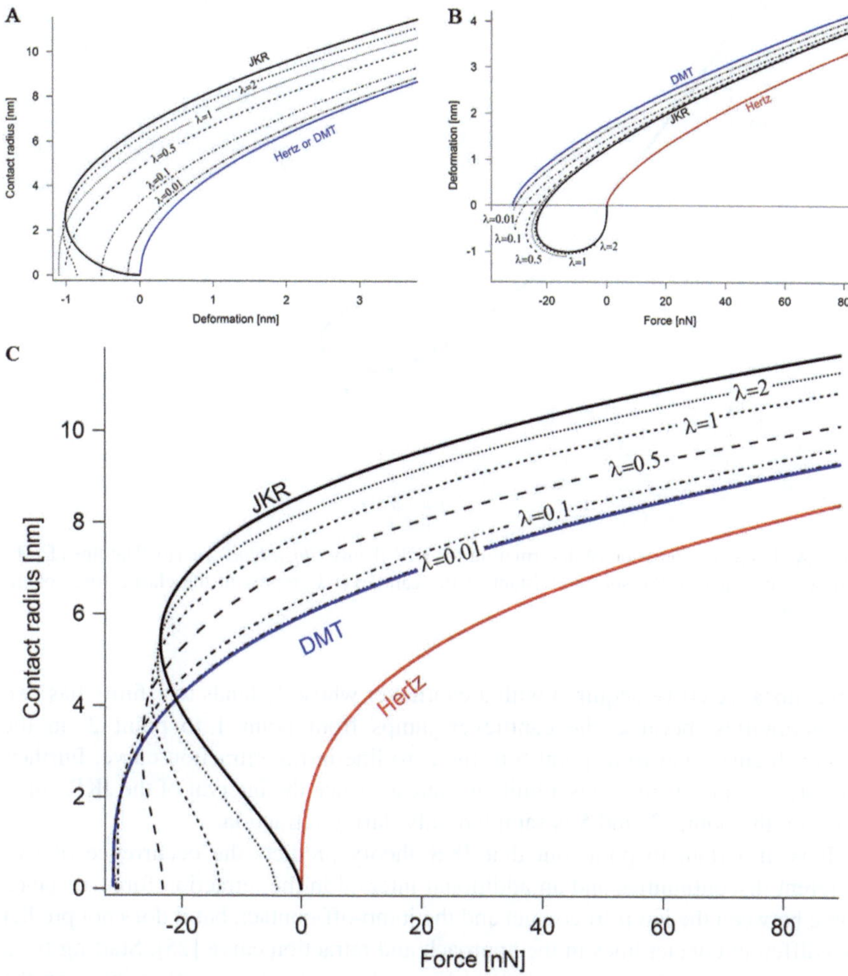

Fig. 1.3 (**A**) Contact radius as a function of the deformation, (**B**) deformation as a function of the force and (**C**) contact radius as a function of the force in Hertz (*red*), DMT (*blue*), JKR (*black*) and in Maugis theory for $\lambda = 0.01$, 0.1, 0.5, 1 and 2 (*dashed lines*). The maximum adhesion force in DMT theory is $2\pi RW \approx 31$ nN and the corresponding contact radius is zero. The maximum adhesion force in JKR theory is $3/2\pi RW \approx 23$ nN and the corresponding contact radius is $a = a_{min} \approx 5.4$ nm

1, corresponding to the jump-to-contact distance Z_{jtc}, the cantilever jumps to the point 2; during the retraction, at point 4, corresponding to the jump-off-contact distance Z_{joc}, the cantilever jumps to the zero line. The essential difference between this hysteresis and the one illustrated in Sect. 1.2 is that the hysteresis due to JKR theory cannot be avoided, even with an infinitely stiff cantilever, whose elastic force is shown through dashed lines. Since there are two values of the deformation for negative loads and the first derivative of $D(F)$ at point 5 tends to infinity, even a

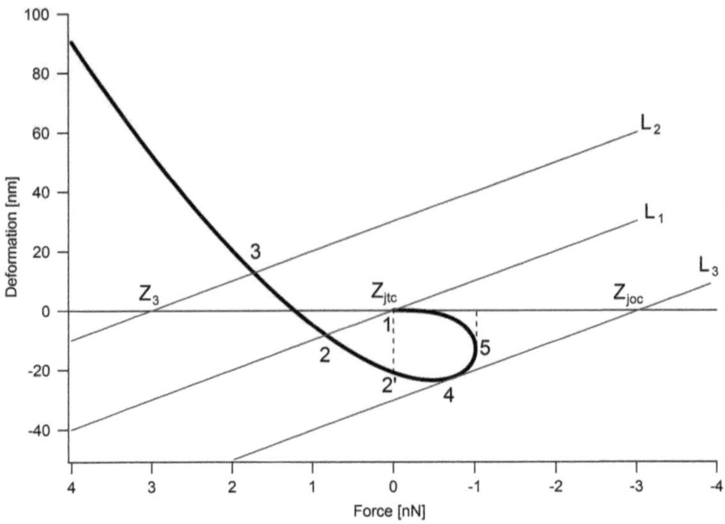

Fig. 1.4 Force as a function of deformation for JKR theory (*thick black curve*). The lines L_1–L_3 with a slope equal to the spring constant of the cantilever k_c represent the elastic force of the cantilever

force–distance curve acquired with a cantilever whose k_c tends to infinity has two discontinuities, because the cantilever jumps from point 1 to point 2′ in the approach curve and from point 5 to the zero line in the retraction curve. Further, the force–distance curve has a finite hysteresis, since the interval of the JKR curve between the points 2′ and 5 is sampled only during retraction.

It is important to point out that JKR theory predicts the occurrence of two different discontinuities and an additional interval in the retraction force–distance curve between the jump-to-contact and the jump-off-contact, but it does not predict two different contact lines in the approach and retraction curve [25]. Starting from the jump-to-contact (points 2 or 2′ in Fig. 1.4), the contact is reversible and the contact lines are exactly the same. A hysteresis of the contact lines, implying energy dissipation, can be engendered only by plastic deformation or by deformations which are recovered with a finite characteristic time, i.e. when the sample has viscoelastic behaviour, as explained in Sects. 1.8 and 1.9.

Now we turn back to Maugis theory. For the calculation of the Maugis curves in Fig. 1.3, given a certain value of the parameter λ, for each value of $\bar{A}$, the parameter m_M has been computed numerically from Eq. (1.36). Putting this value in Eqs. (1.33) and (1.34) yields the deformation and the load as a function of the contact radius, $D(a)$ and $F(a)$, or also the deformation as a function of the load, $D(F)$. This curve can be used to measure the values of W and E_{tot}, but these values must be known since the very beginning, in order to calculate λ!

The Maugis theory shows that it is not possible to measure both the surface energy W and Young's modulus E only with a force–distance curve, since, other than in the DMT and JKR theory, the maximum adhesion force depends also on E and cannot be used to determine W without knowing E. On the other side, the function $D(F)$ depends also on W and cannot be used to determine E without knowing W.

Exact values of E and W can be calculated only in the limiting cases corresponding to Hertz, DMT and JKR theory, since in these cases the surface energy is zero or can be determined from the maximum adhesion, and thus the deformation–load curve does not depend on W.

1.4.4 Oliver and Pharr Theory

The theory of Oliver and Pharr [26] was originally developed for the analysis of indentations performed with nanoindenter.

This theory is based on the assumption that, at the beginning of the unloading process, the indented material responds elastically.

Figure 1.5 shows a schematic vertical cross section of an indented material. Several quantities are defined therein. In measurements with nanoindenters, indentations are traditionally denoted with the symbol "h", which corresponds to the symbol "D", traditionally employed for deformations in AFM experiments. Furthermore, the load is usually denoted with "P" instead of "F", and the reduced modulus is defined as $1/E_{\text{tot}} = \left[(1-\nu^2)/E + \left(1-\nu_t^2\right)/E_t\right]$, i.e. without a factor 3/4. The definition in Eq. (1.12) has been maintained in this book; hence, the following equations differ by the factor 3/4 from formulas used in most articles.

The indentation h can be written as

$$h = h_c + h_s, \tag{1.41}$$

where h_c is the vertical distance along which indenter and material have been in contact (contact depth) and h_s is the indentation at the perimeter of the contact.

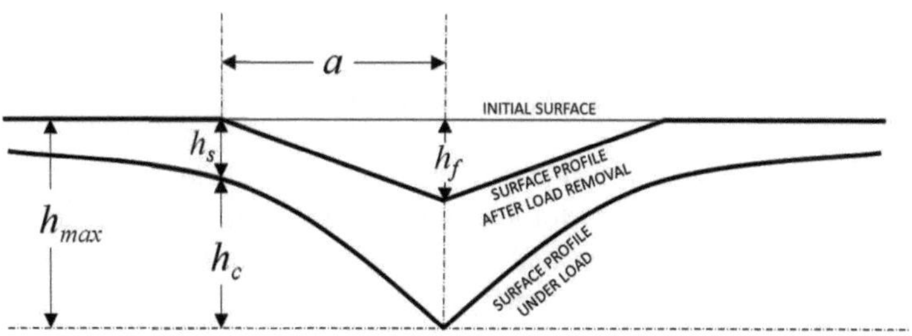

Fig. 1.5 Schematic representation of the vertical cross section of an indented material. The following quantities are shown: the contact radius a, the maximum indentation h_{max}, the indentation at the perimeter of the contact h_s, the contact depth h_c and the indentation after load removal or residual indentation h_f

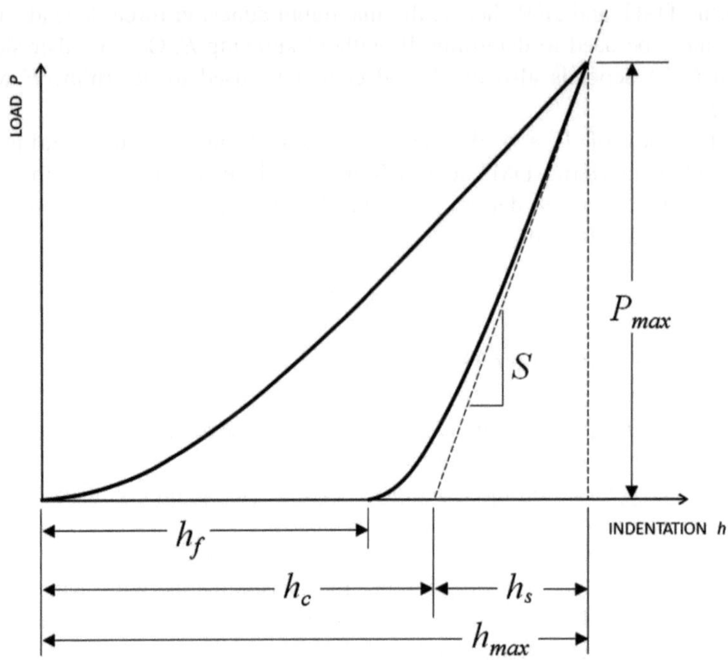

Fig. 1.6 Schematic representation of the load–indentation curve corresponding to Fig. 1.5. The following quantities are shown: the maximum load P_{max}, the maximum indentation h_{max}, the residual indentation h_f, the contact depth h_c, the elastic recovery at maximum load h_s and the initial unloading stiffness S

After reaching the maximum indentation h_{max} at the maximum load P_{max}, upon unloading, elastic deformations are recovered and, when the load is removed, the residual or final indentation is h_f.

The load–indentation curve corresponding to the indentation depicted in Fig. 1.5 is shown in Fig. 1.6.

The analysis of the curve begins by rewriting Eq. (1.11) in the form [27]

$$S = \left.\frac{dP}{dh}\right|_{h_{max}} = \frac{3}{2}\sqrt{\frac{A_{max}}{\pi}}E_{tot}, \tag{1.42}$$

where A_{max} is the cross-sectional area at maximum indentation.

Equation (1.42) is valid only for axisymmetric indenters. If necessary, the right term is multiplied by a dimensionless correction factor β, which accounts for deviations due to the lack of symmetry of the indenter. The value of β is 1.012 for a square-based Vickers indenter and 1.034 for a triangular Berkovich punch [28, 29].

Assuming that the indenter geometry can be described by a function $F(h)$ relating its cross-sectional area to the distance h from the apex, A can be written as $A = F(h_c)$.

Via Eq. (1.41), h_c can be estimated, if the elastic recovery at maximum load, h_s, is known.

For a flat punch is $h_s = P_{max}/S$; for other indenter geometries, the following relation is valid:

$$h_s = \varepsilon \frac{P_{max}}{S}. \tag{1.43}$$

The factor ε depends on the tip shape and spans from 0.72 (conical indenter) to 1 (flat punch). For a paraboloid and for a Berkovich pyramid, it assumes the value $\varepsilon = 0.75$ [27].

In order to determine the indenter geometry and to choose the appropriate value of the factor ε, the unloading curve is fitted with a power law in the form

$$P = \alpha(h - h_f)^n. \tag{1.44}$$

The value of the exponent n, as already explained, is given by the indenter geometry: $n = 2$ for a cone, $n = 1.5$ for a paraboloid or a sphere and $n = 1$ for a flat cylinder.

Once h_c is known, $A = F(h_c)$ can be calculated for different geometries. For axisymmetric geometries A is proportional to h_c^2. The proportionality factor C depends only on the indenter geometry; its value is 24.5 for a Berkovich pyramid [27].

Hence, for the contact area and the reduced modulus, following general relations can be written:

$$A = Ch_c^2 = C\left(h_{max} - \varepsilon \frac{P_{max}}{S}\right)^2, \tag{1.45}$$

$$E_{tot} = \frac{2}{3\beta}\sqrt{\frac{\pi}{C}} \frac{S^2}{(Sh_{max} - \varepsilon P_{max})}. \tag{1.46}$$

The theory of Oliver and Pharr is a semiempirical theory based on the assumption that plastic deformations and adhesion can be neglected when considering the very beginning of the unloading curve and that the factors β and ε depend only on the tip geometry and can be determined through a measurement with a reference sample.

However, the portion of the curve used for the analysis is the unloading curve, i.e. the part of the curve where time-depending phenomena are more likely to occur [30]. The influence of viscoelastic phenomena on loading–unloading cycles depends also on the shape of the indenter tip [31]. In particular, when indenting a polymer with a Berkovich tip, due to the edges of the tip, the contact area deviates from the function described in Eq. (1.45). Moreover, the theory of Oliver and Pharr neglects the neck formation and the resulting additional adhesion, a phenomenon which is likely to occur with compliant, highly adhesive samples.

Thus, the assumptions of Oliver and Pharr theory are mostly correct for hard and stiff materials with small elastic deformations, but severe errors occur in case of soft and compliant materials with viscoelastic behaviour, such as polymers [31, 32].

1.5 Brief Review of Surface Forces

Although surface forces do not influence the mechanical properties of samples, they determine the shape of force–distance curves before contact and the adhesion between tip and sample. The shape before contact plays a major role when measuring the distance at which contact is established, i.e. when measuring the origin of the deformation–force curve. The tip–sample adhesion is crucial for the choice of an elastic continuum theory, as illustrated in Sect. 1.4. Accordingly, a brief review of surface forces is outlined in this section. A complete theoretical and experimental review of surface forces can be found in [11, 12, 33].

When force–distance curves are acquired in air, which is the common case in studies of mechanical properties, a thin water layer adsorbed on the sample surface exerts a *meniscus* or *capillary force* [34]. Due to the high surface tension of water, the meniscus force exerted by water is a strong attractive and adhesive force, much larger than any other surface force. It leads to pronounced discontinuities in the force–distance curve. A pronounced jump-to-contact makes it easier to identify the contact point. On the other side, it must be kept in mind that the capillary force is an interaction between the tip and the water film on the sample surface and not between tip and sample and that the capillary force influences the deformation of the water film and not of the underlying sample. Hence, if the adhesion between tip and sample is estimated from the jump-off-contact, it is overestimated, and the modulus calculated with DMT or JKR theory is lower than the real modulus. On the other hand, since the meniscus force "hides" the tip–sample interaction, there is no possibility to estimate it.

In order to eliminate the meniscus force, it is necessary to acquire force–distance curves in dry atmosphere (typically with humidity smaller than 10 %) [35] or to eliminate the air–water interface by dipping the tip–sample system in water or in other liquids [36].

When working in a medium, other surface forces become relevant. The only surface force present in all systems is the *van der Waals force*.

The van der Waals force between a sphere of radius R and a plane surface at the distance D is given by

$$F = -\frac{A_H R}{6D^2}. \tag{1.47}$$

A_H is the Hamaker constant, given approximately by the sum of two terms. The first term is

$$(A_H)_{v=0} = \frac{3}{4} k_B T \frac{\varepsilon_1 - \varepsilon_3}{\varepsilon_1 + \varepsilon_3} \frac{\varepsilon_2 - \varepsilon_3}{\varepsilon_2 + \varepsilon_3}, \tag{1.48}$$

where k_B is the Boltzmann constant, T the temperature and ε the dielectric permittivity (the indexes 1 and 2 denote the interacting bodies; the index 3 denotes the medium).

The second term is

$$(A_H)_{v>0} = \frac{3hv_e}{8\sqrt{2}} \frac{\left(n_1^2 - n_3^2\right)\left(n_2^2 - n_3^2\right)}{\sqrt{|n_1^2 - n_3^2|}\sqrt{|n_2^2 - n_3^2|}\left(\sqrt{|n_1^2 - n_3^2|} + \sqrt{|n_2^2 - n_3^2|}\right)}, \tag{1.49}$$

where h is the Planck constant, v_e the average absorption frequency and n the refractive index. Whenever $hv_e > k_B T$, the second term dominates the interaction.

The sign of the van der Waals force depends on the ratio of the dielectric permittivities or of the refractive indexes, depending on the dominating term. The van der Waals force between two identical materials is always attractive. The force between two different materials is repulsive whenever ε_3 or n_3 is intermediate between ε_1 and ε_2 or n_1 and n_2, respectively. In air or in vacuum, the van der Waals force is always attractive and is always stronger than in a medium.

A repulsive van der Waals force can "hide" the jump-to-contact, so that it is difficult to determine at which point the contact between tip and sample is established.

Another force mostly present in water is the *double-layer force*, due to the charging of surfaces, for example via dissociation of surface groups or adsorption of ions in solution. The surface charge is balanced by a distribution of counterions in the medium. In order to approach two surfaces in a medium with a distribution of counterions and to compel them to reduce their mutual distance and the distance to the surface, work is needed.

Butt [37] has computed the double-layer force between a flat sample with charge density σ and a spherical tip with radius R and charge density σ_T:

$$F = \frac{2\pi R \lambda_D}{\varepsilon \varepsilon_0} \left[(\sigma^2 + \sigma_T^2) \exp\left(-\frac{2D}{\lambda_D}\right) + 2\sigma\sigma_T \exp\left(-\frac{D}{\lambda_D}\right) \right]. \tag{1.50}$$

In this equation, λ_D is the Debye length, i.e. the specific decay length of the potential and of the force, given by

$$\lambda_D = \sqrt{\frac{\varepsilon \varepsilon_0 k_B T}{e^2 \sum c_i^0 \zeta_i^2}}, \tag{1.51}$$

where c_i and ζ_i are the number density and the valence of the ions in solution.

Equation (1.50) is valid for $R > \lambda_D$. The second term of the force dominates only at large distances when $\sigma_T \approx \sigma_S$; in all other cases, i.e. at short distances and for $\sigma_T \ll \sigma_S$, the first term dominates. When the first term dominates and also when the second term is larger than the first, but the surface charges have the same sign, the double-layer force is repulsive.

As a consequence, the double-layer force reduces strongly the tip–sample adhesion, and finding the point of contact becomes very difficult, also because of the exponential decay of the force [38, 39].

Another noteworthy effect in water is the hydrophobic meniscus, i.e. the strong adhesion force exerted by hydrophobic surfaces in water, due to the fact that hydrophobic samples “prefer” to be in contact with the tip rather than with water [36]. Due to this effect, when measuring the mechanical properties of hydrophobic samples and polymers, working in a medium, instead of decreasing the adhesion, increases it. This may lead to problems in the choice of the elastic continuum theory and to errors in the theoretical modelling of the experimental results.

The theory taking into account both van der Waals and double-layer force is called DLVO (Derjaguin–Landau–Verwey–Overbeek) theory [40, 41]. DLVO theory cannot describe the interaction of two surfaces approaching at distances smaller than few nanometres, because the continuum theories of van der Waals and double-layer force are not valid at small separations and because other forces arise.

The most important non-DLVO forces are the *solvation force* [42], the *hydration force* [43] and the *hydrophobic force* [44].

The solvation force is present whenever liquid molecules are compelled to order in almost discrete layers between two solid surfaces. In this case the density of the liquid follows an oscillatory profile, engendering a force, also described by an oscillatory function:

$$F = -k_B T \rho_\infty \frac{2\pi R D_m}{\sqrt{4\pi^2 + 1}} \cos \frac{2\pi D}{D_m} \exp\left(-\frac{D}{D_m}\right), \tag{1.52}$$

where ρ_∞ is the density of the liquid at the interface when the second surface is infinitely distant and D_m is the molecular diameter. Equation (1.52) is valid for a parabolic tip and a plane.

The hydration force arises from modifications of the H-bonding network of water between hydrophilic surfaces or when hydrated cations in solution bind to a negatively charged surface at high salt concentration. The hydration force increases with the hydration number and decays exponentially with the distance.

The hydrophobic force is probably due to modifications of the water network in the presence of hydrophobic surfaces and can be empirically fitted with an exponential function.

Non-DLVO forces are present only in specific systems: liquids with spherical molecules for the solvation force, water solutions with high electrolyte concentration or hydrophilic surfaces with small contact angle for the hydration force and hydrophobic surfaces with large contact angle for the hydrophobic force. Moreover, they act only at very small separations. Hence, they are not relevant factors for measurements of mechanical properties.

Another category of forces must be considered when studying the mechanical properties of polymer samples. The forces of this category are present between spatially diffuse surfaces, i.e. surfaces with thermally mobile groups like polymer chains dangling out of the surface into the solution.

When another surface is approached, the chains are confined in a smaller volume and a repulsive entropic force arises, called *steric force*. The steric force depends primarily on the way polymer chains are adsorbed, i.e. whether physisorbed or chemisorbed, on the extension of the polymer chains in solution and on the covering of the surfaces. When the polymer is chemisorbed, the covering is expressed in terms of the grafting density Γ, i.e. the number of grafted chains per unit area.

The extension of the polymer chains in solution depends in turn on the size of the polymer chains but also on the interactions between monomers and solvent, i.e. on the quality of the solvent.

In an ideal or θ-solvent, i.e. when the interactions between the monomers in the solvent can be neglected, the size of the chains is given by the unperturbed gyration radius of the polymer $R_g = l\sqrt{n_m/6}$, where n_m is the number of monomers in a chain and l the length of a monomer.

In a real solvent, the size of the chains is given by Flory radius $R_F = \alpha R_g$. The factor α exceeds unity in a good solvent, where the monomers "prefer" to be in contact with the solvent, and $R_F = ln_m^{3/5}$. In a poor solvent, the monomers attract each other and α is less than unity.

Concerning the covering, it is important to distinguish between low ($<1/R_g^2$) and high covering ($>1/R_g^2$). Because of the possible lateral diffusion, physisorbed polymer brushes have always a low covering. For a physisorbed polymer in a θ-solvent, the extension of the polymer chains in solution, t_f, is about $2R_g$.

If the polymer chains are chemisorbed and $\Gamma \gg 1/R_g^2$, we talk about a polymer brush. In this case $t_f = n_m\sqrt[3]{24\Gamma l^5/\pi^2}$.

For polymer brushes the steric force is given by [12, 45, 46]

$$F = 4\pi RP_0\left[\frac{2t_f}{D} + \left(\frac{D}{2t_f}\right)^2 - \frac{1}{5}\left(\frac{D}{2t_f}\right)^5 - \frac{9}{5}\right] \tag{1.53}$$

with $P_0 = \frac{k_B T n_m}{2}\sqrt[3]{\frac{\pi^2 l^4 \Gamma^5}{12}}$.

For low grafting density, the steric force per unit area in a good solvent is given by [12, 47]

$$P = \frac{k_B T\Gamma}{x}\left(\frac{2\pi^2 R_g^2}{x^2} - 1\right) \quad \text{for } x < 3\sqrt{2}R_g, \tag{1.54}$$

$$P = \frac{k_B T\Gamma x}{R_g^2}\exp\left(-\frac{x^2}{4R_g^2}\right) \quad \text{for } x > 3\sqrt{2}R_g, \tag{1.55}$$

where x is the gap between the surfaces.

When the surfaces are closer than R_g, the chains are pushed out of the gap, resulting in a reduced polymer concentration which engenders an attractive force, the so-called depletion force, given by [48]

$$F = \pi \frac{\mu}{V_m} (D + 2R)(D - 2t_f), \tag{1.56}$$

in which μ is the chemical potential of the solvent and V_m its molecular volume.

A third force between a polymer-covered surface (sample) and a second bare surface (tip), the *bridging force*, depends on the polymer–surface interactions. If there are free binding sites on the opposite surface, some chains will form bridges between the two surfaces giving rise to an attractive force [49]. The bridging force decays roughly exponentially with the distance, with a decay length close to R_g. The formed bridges also exert a strong adhesion upon retraction of the second surface.

These three forces can change strongly the tip–sample adhesion; moreover, the steric force, being repulsive, can make it difficult to determine the contact point.

1.6 Mechanical Properties of Polymers

Other than most solids, whose elastic modulus stays constant, as long as the material does not melt, the elastic modulus of amorphous polymers depends on the temperature. Additionally to the melting temperature T_m, amorphous polymers present also a second typical temperature, the glass transition temperature T_g.

The glass transition temperature [13, 50, 51] is a specific temperature, at which pure or partly amorphous polymers go from the glassy over to the rubbery state and therewith change from a brittle to a highly viscous and flexible material.

Other than melting, which is a first-order transition, the transition from the glassy to the rubbery state presents some characteristics of a second-order transition. The difference between a first- and a second-order transition can be understood by considering the thermodynamics of the transition and in particular Gibbs free energy G:

$$G = U - TS + PV. \tag{1.57}$$

Here, U is the internal energy of the system, T the temperature, S the entropy, P the pressure and V the volume.

In a first-order transition such as melting, the first derivatives of Gibbs free energy G, e.g. the volume V and the entropy S, are discontinuous. In a second-order transition, the first derivatives of G are continuous. Yet, the second derivatives of G, such as the thermal expansion coefficient $\alpha_V = \mathrm{d}V/\mathrm{d}T$, are discontinuous.

Figure 1.7 shows two volume–temperature curves for a crystalline solid (dashed lines) and for a polymer (continuous lines). By increasing the temperature, the volume of the crystalline solid increases (A′B′), and finally the solid passes in the

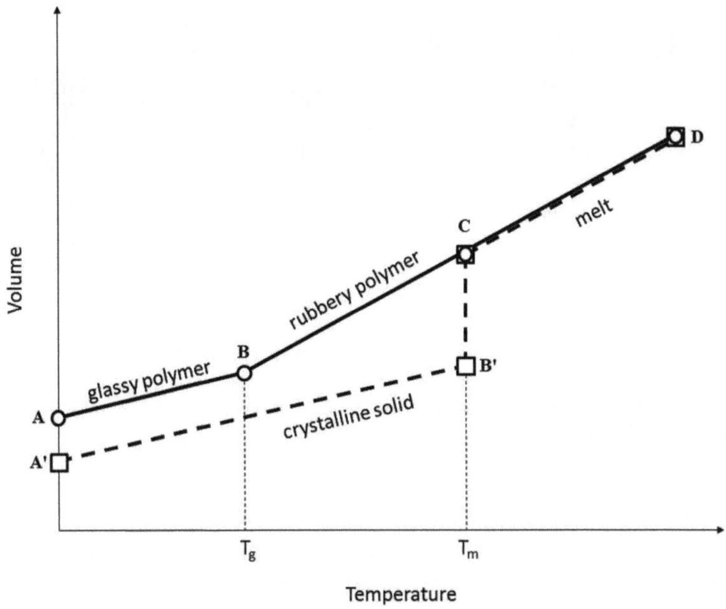

Fig. 1.7 Volume–temperature curves for a crystalline solid (A′B′) going over to the liquid phase (CD) at the melting temperature T_m (*dashed lines*) and for a polymer in the amorphous glassy state (AB) going over at first to the rubbery state (BC) at the glass transition temperature T_g and finally to the liquid phase (CD) at the melting temperature T_m (*continuous lines*)

liquid state at the melting temperature T_m. This transition, as already said, is accompanied by an abrupt increase in volume. The slope of the volume–temperature curve is the thermal expansion coefficient α_V, which is higher in the liquid phase than in the solid phase.

Also the volume of the polymer increases with increasing temperature (AB). Yet, the polymer undergoes a transformation from the glassy to the rubbery state at the glass transition temperature T_g. This transformation is not accompanied by a discontinuity.

The initial volume occupied by the polymer in the glassy state is larger than that occupied by a corresponding crystalline solid, due to lack of ordering in the system. This excess volume is called the free volume V_f.

Most polymers show in addition to the glass transition two other transitions, the gamma and beta transition [52]. All thermal transitions in polymers can be described in terms of either free volume changes or relaxation times.

A simple approach to the concept of free volume, which is useful in explaining the dynamic mechanical properties of polymers, is the crankshaft model, where polymer chains are considered as a collection of mobile segments, which have a certain free volume available, as illustrated schematically in Fig. 1.8.

When the temperature is increased, the free volume increases too, and the segments can perform various movements. The movements requiring less free

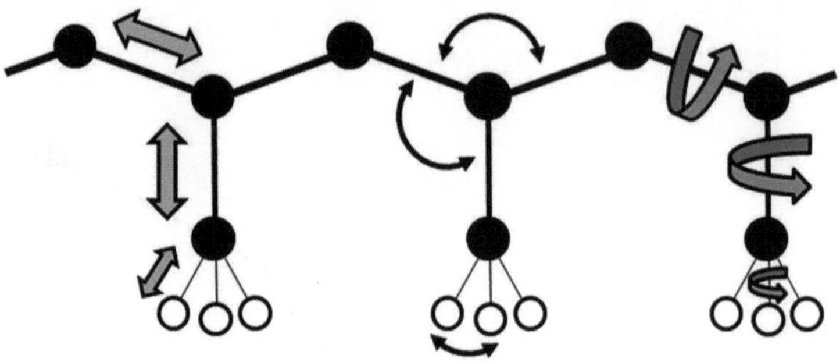

Fig. 1.8 Schematic illustration of a polymer molecule with a part of the main chain and three side groups. The possible movements of parts of the molecule are indicated for parts of the side group, the whole side group and parts of the main chain: stretching (*grey arrows*), bending (*black arrows*) and rotating ("3D"-*arrows*)

volume (and less energy) are localized bond stretching, bending and rotations of parts of the side chains. Such movements engender the gamma transition at the temperature T_γ. With a further increase of the temperature, whole side chains and groups of some atoms in the main chain can also stretch, bend or rotate. This transition is the beta transition. Finally, when large segments of the main chain have enough space and energy to move, the glass transition occurs at the temperature T_g. The last transition at the melting temperature T_m occurs when the free volume is so large that whole chains can slide past each other and the material flows.

For each transition an activation energy E_a can be defined [52], which must be overcome by a certain group of atoms in order to perform the respective movements. The relaxation time τ_r, i.e. the typical time, in which a transition takes place, can be expressed in terms of the activation energy and of the thermal energy of the group of atoms (k_BT):

$$\tau_r = A\exp\frac{E_a}{k_BT}, \tag{1.58}$$

where A is a proportionality factor.

The glass transition temperature depends mainly on the chemical structure and mobility of the polymer chains. The molecular weight, the stiffness of the chains, intermolecular forces, cross-linking and ramification of the side chains influence the molecular mobility and thus the glass transition temperature [53].

The dependence of the glass transition temperature of a polymer on its mean weight M_n is of particular interest [13, 50]. When the temperature of a polymer with molar mass M_n and glass transition temperature T_g^M reaches the glass transition temperature of a polymer with infinite molar mass (T_g^∞), an excess free volume v_f can be assigned to each chain end. With N_A the Avogadro number and ρ the density, the excess free volume is given by $V_f = 2\rho v_f N_A / M_n$. The excess free volume can be expressed also as the thermal expansion between T_g^M and T_g^∞, i.e.

$V_f = \alpha_V \left[T_g^{\infty} - T_g^M \right]$, where α_V is the thermal expansion coefficient. From both expressions of the excess free volume, Fox–Flory relation is obtained [54]:

$$T_g^M = T_g^{\infty} - \frac{2\rho v_f N_A}{\alpha_V M_n} \Rightarrow T_g^M = T_g^{\infty} - \frac{K}{M_n}, \tag{1.59}$$

where K is a material-specific constant.

As anticipated at the beginning of the section, the glass transition temperature has a strong influence also on the mechanical properties of a polymer and in particular on Young's modulus.

At temperatures below T_g, a polymer is in its glassy state, where the value of the elastic modulus is typically in the range 1–10 GPa. At temperatures above T_g, a polymer is in its rubbery state, and due to the increased free volume and the enhanced chain mobility, its modulus is commonly in the range 1–100 MPa.

In a temperature interval around T_g, whose width depends on the specific polymer, the polymer is in the transition region; it has an intermediate modulus and viscoelastic properties.

1.7 Time–Temperature Superposition Principle and Williams–Landel–Ferry Equation

Equation (1.58) shows that time and temperature have an equivalent effect on a transition. This equivalence is enounced in the time–temperature superposition principle [13, 51, 55], which says that the effect of a time extension (or frequency decrease) is equal to that of a temperature increase for several polymer properties, e.g. mechanical and dielectric properties. Given a certain mechanical quantity of a polymer, e.g. the elastic modulus E, the function $E(T_{\text{ref}}, \nu)$, with T_{ref} a given temperature and ν the frequency, has a similar shape as other isotherms at neighbouring temperatures. Hence, it is possible to record several isotherms $E(T_i, \nu)$ at different temperatures and frequencies and then shift the isotherms till they overlap the reference isotherm at T_{ref}. Shifting all isotherms yields the master curve, whose frequency and temperature range is wider than the range of the experimental frequencies and temperatures. Figure 1.9 shows the isotherms of the elastic modulus E of poly(n-butyl methacrylate) measured with dynamic mechanical analysis.

By shifting the isotherms till they overlap with the reference isotherm at $T_{\text{ref}} = 23$ °C, the master curve of the elastic modulus of poly(n-butyl methacrylate) as function of frequency, shown in Fig. 1.10, is obtained.

The shift factors of the isotherms, $\log a_T$, are related to the temperature through the Williams–Landel–Ferry (WLF) equation [56] or through the Arrhenius equation.

The first equation can be used in the temperature range between T_g and $T_g + 100$ K and reads:

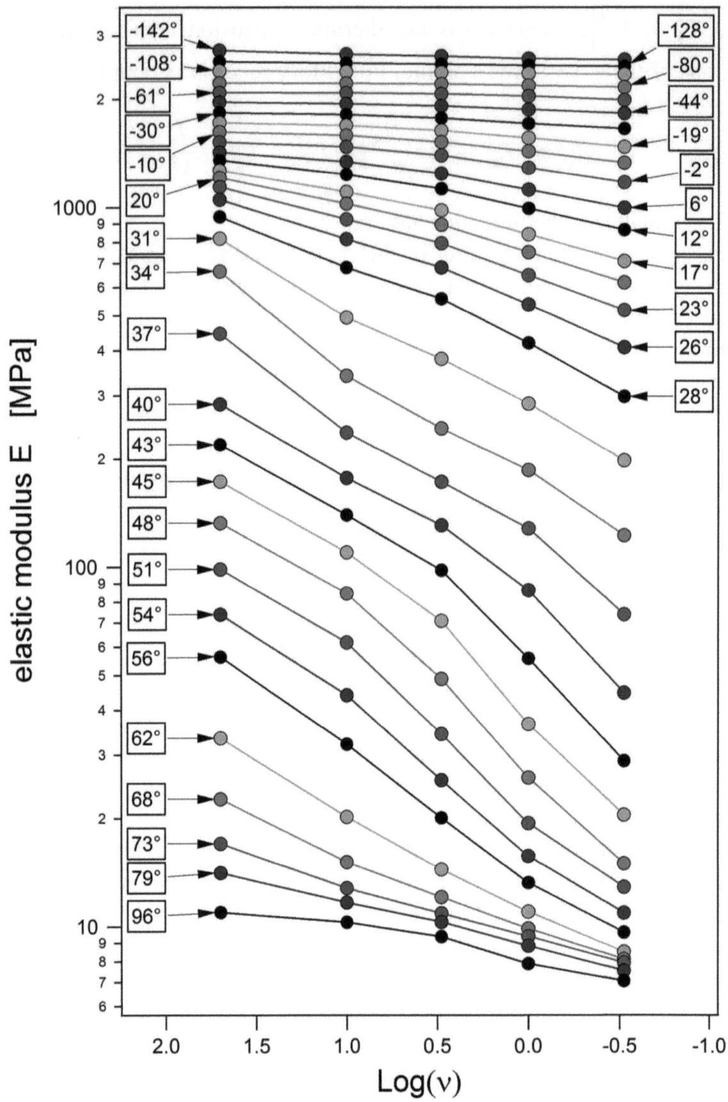

Fig. 1.9 Isotherms of the elastic modulus E of poly(n-butyl methacrylate) vs. the logarithm of the frequency. The temperatures are indicated in the tags

$$\log a_T = \frac{-C_1(T - T_{\text{ref}})}{C_2 + (T - T_{\text{ref}})}, \tag{1.60}$$

where C_1 and C_2 are constants depending on the choice of the reference temperature. If a different reference temperature T'_{ref} is chosen, the relation between the constants is given by

$$C_2' = C_2 + T_{\text{ref}}' - T_{\text{ref}}, \quad (1.61)$$

$$C_1' = \frac{C_1 C_2}{C_2'}. \quad (1.62)$$

If $T_{\text{ref}} = T_g$, the values of the constants are for most polymers $C_1 = 17.44$ and $C_2 = 51.6\ °\text{C}$.

The Arrhenius equation is used when $T < T_g$:

$$\ln a_T = \frac{E_a}{R_g}\left(\frac{1}{T} - \frac{1}{T_{\text{ref}}}\right) + \ln A, \quad (1.63)$$

where E_a is the activation energy of the respective transition and R_g is the universal gas constant (8.31 J/mol K).

Since shift factors are frequencies, WLF and Arrhenius equations permit to establish the equivalence between temperature and frequency and/or time. Such equivalence is the same for all properties obeying the time–temperature superposition principle, including mechanical and dielectric properties. Hence, Fig. 1.10 shows not only the dependence of the modulus of poly(*n*-butyl methacrylate) on the frequency but also on the temperature. At high frequencies or low temperatures ($\log(\nu) > 10$ or $T < 0\ °\text{C}$), the polymer is in its glassy state and has a modulus between

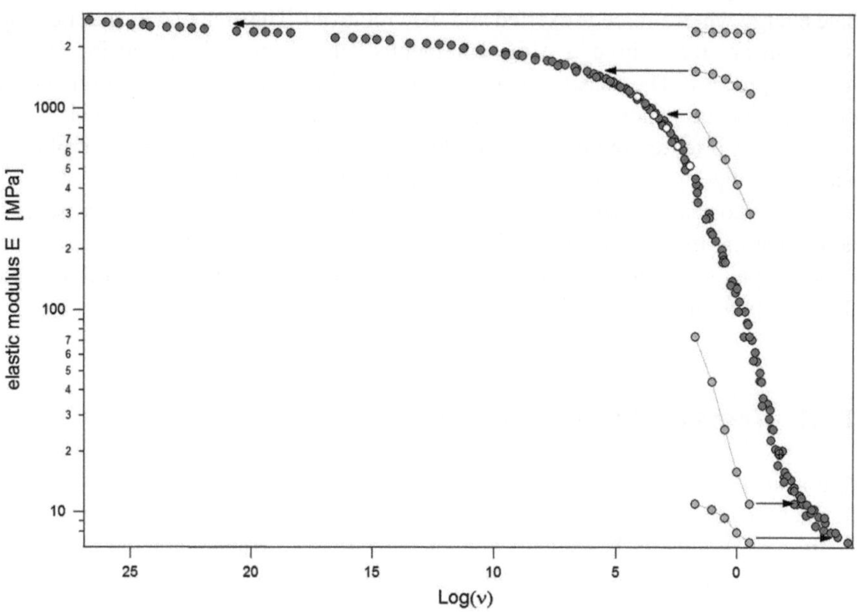

Fig. 1.10 Generation of the master curve through shift of the isotherms in Fig. 1.9 following the time–temperature superposition principle. The reference temperature (*empty circles*) is $T_{\text{ref}} = 23\ °\text{C}$. Five isotherms are shown in light grey also before being shifted. The *arrows* indicate the shift factors $\log a_T$

1 and 2 GPa, slightly decreasing with increasing temperature or decreasing frequency. At low frequencies or high temperatures ($\log(\nu) < 1$ or $T > 60$ °C), the polymer is in its rubbery state and has a modulus smaller than 10 MPa, slightly decreasing with increasing temperature or decreasing frequency. In the transition region, the modulus decreases very steeply; the transition region is 60–70 °C wide, corresponding to approximately 10 decades of the frequency.

1.8 Viscoelasticity and Loss Elastic Modulus

A polymer in the transition region and in the rubbery state has viscoelastic properties. Viscoelasticity [13, 51] is a time- and temperature-dependent elasticity. Viscoelasticity is characterized by a partially elastic and partially viscous behaviour. As already seen, for an elastic material, deformations are proportional to the stress σ and do not depend on the stress rate $d\sigma/dt$. For viscous liquids, deformations are proportional to the stress rate and are independent of the stress.

For a linear viscoelastic material, deformations (and consequently the modulus) depend both on the stress and on the stress rate. Hence, when a material exhibits viscoelastic behaviour, the deformation depends on the loading history, and the shear–stress curve on a viscoelastic material presents hysteretic loops depending on the loading pattern. Other than pure elastic materials, viscoelastic materials dissipate energy when a force is exerted and then removed. The area of a hysteretic loop in the shear–stress curve is a measurement of the energy dissipated in a given loading process. The dissipated energy results in a plastic deformation, i.e. a deformation, in which the deformed material cannot regain its original shape immediately after removing the force. Plastic deformations of a polymer in rubbery state are not permanent, i.e. they relax with a certain time delay.

In order to characterize the viscoelastic behaviour of a polymer, it is necessary to determine the modulus as a function of time, i.e. for a particular loading pattern. Loading patterns can be divided in two large categories, transient patterns and dynamic patterns.

1.8.1 Transient Loading Patterns

A transient loading pattern is a suddenly application of a stress. An example of transient loading pattern is creep.

In creep, a stress σ_0 is suddenly applied at the time $t = 0$ and maintained constant for a time t_L, as illustrated schematically in Fig. 1.11A.

In case of creep loading, if the material has an ideally elastic behaviour, the strain follows the stress without delay, i.e. it attains suddenly the value ε_0 at $t = 0$ and becomes again zero at $t = t_L$, as illustrated in Fig. 1.11B.

For a material with viscoelastic behaviour, the time-dependent strain $\varepsilon(t)$ resulting from a stress σ_0 applied at $t = t_0$ is given by [13]

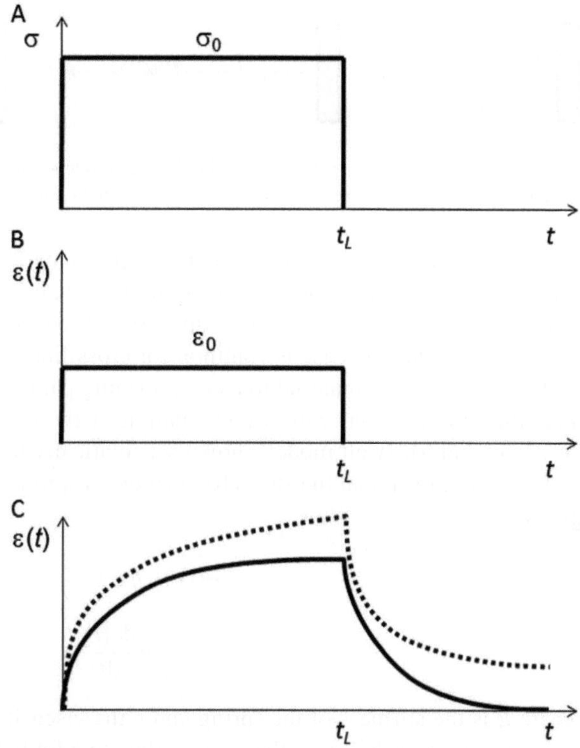

Fig. 1.11 Stress σ_0 applied for a time t_L (**A**), strain response of an ideal elastic material (**B**) and strain response of a material with viscoelastic behaviour (**C**) and with (*continuous line*) or without (*dotted line*) an equilibrium compliance J_∞

$$\varepsilon(t) = \sigma_0 J(t - t_0), \tag{1.64}$$

where $J(t)$ is the time-dependent compliance, i.e. $J(t) = 1/E(t)$.

If additional stresses σ_i are added at the times t_i, their effects on the strain can be generalized following Boltzmann superposition principle:

$$\varepsilon(t) = \sum_{i=0}^{i=\infty} \sigma_i J(t - t_i), \tag{1.65}$$

or, for a continuous stress variation,

$$\varepsilon(t) = \int_0^\infty \sigma(t - s)\frac{\mathrm{d}J(s)}{\mathrm{d}s}\mathrm{d}s. \tag{1.66}$$

Since removing a stress σ_0 is equivalent to applying a stress $-\sigma_0$, the response of a material with viscoelastic behaviour, illustrated in Fig. 1.11C, is given by

$$\varepsilon(t) = \sigma_0[J(t) - J(t - t_L)]. \tag{1.67}$$

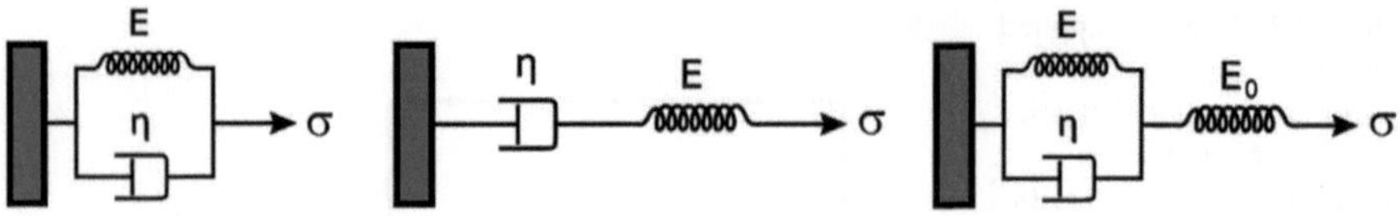

Fig. 1.12 *Left* and *middle panel*: basic rheological models, i.e. Voigt model (*left*) and Maxwell model (*middle*). *Right*: three-parameter or standard linear solid model

The response differs fundamentally if the material possesses an equilibrium compliance $J_\infty = 1/E_\infty$ and a strain equilibrium value $\varepsilon_\infty = \sigma_0 J_\infty$, which has been reached in the time t_L (continuous line), or if there is no equilibrium compliance (dotted line). The first case is common for cross-linked polymers.

The response of a material to a creep loading path can be expressed analytically modelling the material with a combination of the two basic rheological models, i.e. Voigt and Maxwell model, shown schematically in Fig. 1.12.

It is well known that the stress for a Hooke's spring and a Newton's dashpot are given by

$$\sigma(t) = E\varepsilon(t), \tag{1.68}$$

$$\sigma(t) = \eta \frac{\mathrm{d}\varepsilon(t)}{\mathrm{d}t}, \tag{1.69}$$

where E is the stiffness of the spring and η the viscosity of the dashpot.

In the Voigt model, since the strain is the same for both elements and the stresses are additive, the total stress is given by

$$\sigma(t) = \eta \frac{\mathrm{d}\varepsilon(t)}{\mathrm{d}t} + E\varepsilon(t). \tag{1.70}$$

The response to a step stress σ_0 is

$$\varepsilon(t) = J\sigma_0 \left[1 - \exp\left(-\frac{t}{\tau}\right)\right], \tag{1.71}$$

where $\tau = \eta/E$ is the retardation time and $J = 1/E$.

In the Maxwell model, since the stress is the same for both elements and the strains are additive, the total strain is given by

$$\frac{\mathrm{d}\varepsilon(t)}{\mathrm{d}t} = J\frac{\mathrm{d}\sigma(t)}{\mathrm{d}t} + \frac{\sigma(t)}{\eta}. \tag{1.72}$$

The response to a step stress σ_0 $(\mathrm{d}\sigma(t)/\mathrm{d}t = 0)$ is

$$\varepsilon(t) = \varepsilon_0 + \frac{\sigma_0}{\eta}t. \tag{1.73}$$

In the three-parameter or standard linear solid (SLS) model, a spring and a Voigt unit are in series; hence the stress is the same for both components and the strains are additive. The total strain is given by

$$\varepsilon(t) = \sigma_0\left\{J_0 + J_V\left[1 - \exp\left(-\frac{t}{\tau}\right)\right]\right\}, \tag{1.74}$$

where $J_0 = 1/E_0$ and $J_V = 1/E_V$ are the compliances of the spring and of the Voigt unit, respectively.

With $J_\infty = J_0 + J_V$, Eq. (1.74) reads

$$\varepsilon(t) = \sigma_0\left[J_\infty - (J_\infty - J_0)\exp\left(-\frac{t}{\tau}\right)\right] = \sigma_0\Phi(t). \tag{1.75}$$

$J_0 = 1/E_0$ and $J_\infty = 1/E_\infty$ are the instantaneous and infinite or equilibrium compliance, and the function $\Phi(t)$ is the creep compliance.

Other than claimed in [57], $\Phi(t)$ is not a sigmoidal function, since its derivative is monotonically decreasing and not a bell function; in other words $\Phi(t)$ has a plateau for $t/\tau \to +\infty$, but not for $t/\tau \to -\infty$. Actually, $\Phi(t)$ goes from the value $1/E_0$ for $t/\tau = 0$ to the value $1/E_\infty$ for $t/\tau \to \infty$ and is a monotonically increasing, concave function. Hence, the response of a viscoelastic material to a step load can be approximated with the response of an elastic material with moduli E_0 and E_∞ for $t \ll \tau$ and $t \gg \tau$, respectively.

When a force–distance curve is acquired, the load is not a step function. In order to calculate the response of a viscoelastic material described by the SLS model during indentation, Johnson [25] has applied the viscoelastic correspondence principle to Hertz theory, assuming that the applied load is proportional to the time, i.e. $F = v_F t$, where v_F is a constant rate.

Analogously to the equation $D^{3/2} = F/\left(E_{\text{tot}}\sqrt{R}\right)$ (Eq. 1.19), the time-dependent deformation is written as

$$D^{3/2}(t) = \frac{1}{\sqrt{R}}\int_0^t \Phi(t-t')\frac{dF(t')}{dt'}dt' = \frac{3v_F(1-\nu^2)\tau}{4\sqrt{R}E_\infty}\left\{\frac{t}{\tau} - \left(1 - \frac{E_\infty}{E_0}\right)\left[1 - \exp\left(-\frac{t}{\tau}\right)\right]\right\}, \tag{1.76}$$

where the factor $1/E_t$ is ignored in the expression of the reduced modulus.

For $t/\tau \to 0$, $1 - \exp(-t/\tau) \cong t/\tau$ and $D^{3/2}(t) = \frac{3(1-\nu^2)}{4}\frac{F}{\sqrt{R}E_0}$, which is Hertz deformation of a material with modulus E_0.

For $t/\tau \to \infty$, $1 - \exp(-t/\tau) \cong 1$ and $D^{3/2}(t) = \frac{3(1-\nu^2)v_F}{4\sqrt{R}E_\infty}\left[t - \tau\left(1 - \frac{E_\infty}{E_0}\right)\right]$, which is Hertz deformation of a material with modulus E_∞, starting at the time $t_0 = \tau(1 - E_\infty/E_0)$.

The intersection of these two limiting lines is $t = \tau$.

Yet, the assumption that $F = v_F t$ does not hold for a force–distance curve. The only parameter that can be controlled during the acquisition of a force–distance curve is the piezo displacement Z, which is usually indeed proportional to the time, i.e. $Z = v_Z t$. Recalling Eq. (1.5), $F = k_c\delta = \frac{k_c k_s}{k_c + k_s} Z = k_{eff} Z$, it is immediately evident that the load is proportional to time only when k_s does not depend on time. Yet, for a viscoelastic material, modulus and stiffness do depend on time. For $t/\tau \to 0$, i.e. in the first part of an indentation or of a Z ramp, the sample behaves as a perfectly elastic material with modulus E_0 and deformations $D = \left(\frac{3(1-\nu^2)F}{4E_0\sqrt{R}}\right)^{2/3}$. For $t/\tau \to \infty$ the sample behaves as a perfectly elastic material with lower modulus E_∞ and larger deformations $D = \left(\frac{3(1-\nu^2)F}{4E_\infty\sqrt{R}}\right)^{2/3}$. The sample displacement Z per unit time is constant and is the sum of the deflection δ per unit time and of the sample deformation D per unit time. As the deformation increases with time, the load F, proportional to δ, decreases with time. Hence, by postulating the proportionality between load and time, we ignore just the phenomenon that we want to study.

1.8.2 Dynamic Loading Patterns

A dynamic loading pattern is the application of a periodically varied stress, i.e. $\sigma = \sigma_0 \sin(\omega t + \varphi)$, where ω is the period of the oscillating deformation, t the time and φ the phase shift. Such periodic stress generates a periodic strain $\varepsilon = \varepsilon_0 \sin(\omega t)$.

In case of a periodic stress, a complex modulus $E^* = E' + iE''$ can be defined, where E' is the storage modulus and E'' the loss modulus:

$$E' = (\sigma_0/\varepsilon_0)\cos\phi, \tag{1.77}$$

$$E'' = (\sigma_0/\varepsilon_0)\sin\phi. \tag{1.78}$$

The storage modulus E' is hence the in-phase component of the complex modulus E^*; the loss modulus E'' is the out-of-phase component. The storage modulus is a measure of the energy stored per cycle, whereas the loss modulus is a measure of the energy dissipated or lost as heat per cycle.

Figure 1.13 shows both moduli together with the loss factor $\tan\delta = E''/E'$ of poly (*n*-butyl methacrylate). The three master curves are the result of a dynamic mechanical measurement (see Sect. 1.7).

Like the storage modulus, also the loss modulus and hence the loss tangent depend on temperature and frequency and change strongly in the transition region around T_g. In particular the loss modulus, which increases before T_g, decreases very steeply in the transition region. Accordingly, the ratio $\tan\delta$, which is almost zero at temperatures far away from T_g, has a peak with a maximum at about 1 in the transition region.

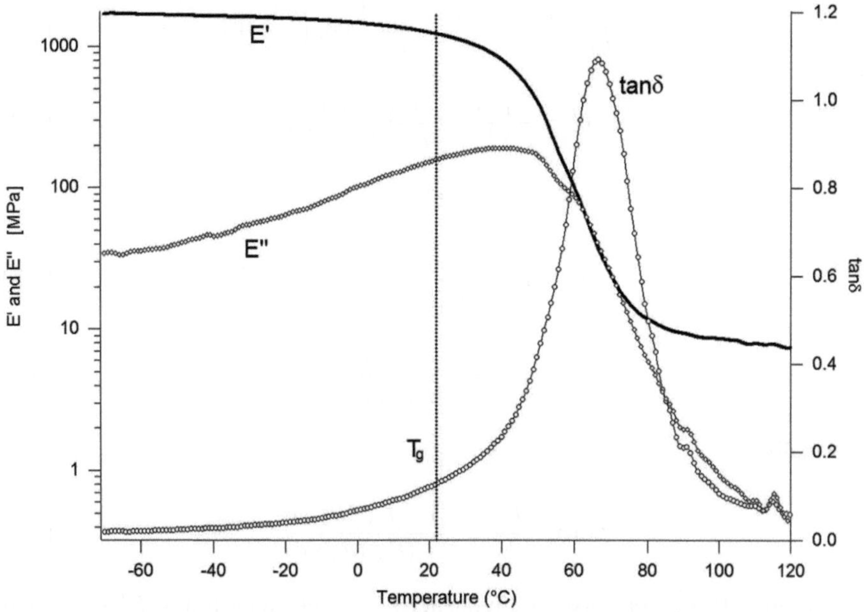

Fig. 1.13 Storage modulus E' (*line*), loss modulus E'' (*diamonds*) and loss factor tanδ (*circles*) of poly(n-butyl methacrylate). The *dotted line* marks the glass transition temperature $T_g = 22$ °C

Figure 1.13 shows also that a definition of the glass transition temperature based on the temperature or frequency dependence of the moduli is rather ambiguous. At least five different temperatures can be chosen: the maxima of E'' and tanδ, the onsets of decrease of E' and E'' and the onset of increase of tanδ (intersections of the linear fits of E', E'' and tanδ in the glassy state and in the transition region). Since the glass transition is actually an interval, it is much more informative to determine the temperature dependence of the moduli in an interval around T_g, as done in Sects. 3.14 and 3.15.

In order to establish a correspondence between transient and dynamic loading patterns, it is necessary to remember that, for a spring, $E^* = E$ and, for a dashpot, $E^* = i\omega\tau E$.

Hence, for a Voigt unit, since the moduli of elements in parallel are additive, $E^* = E + i\omega\tau E$ and

$$J' = \frac{1}{E}\frac{1}{1+\omega^2\tau^2}, \tag{1.79}$$

$$J'' = -\frac{1}{E}\frac{\omega\tau}{1+\omega^2\tau^2}. \tag{1.80}$$

A SLS unit is a Voigt unit in series with a spring; hence the compliances are additive. Remembering that $1/E = 1/E_\infty - 1/E_0$, for a SLS unit is

$$J' = \left(\frac{1}{E_\infty} - \frac{1}{E_0}\right)\frac{1}{1+\omega^2\tau^2} + \frac{1}{E_0} = \frac{1}{1+\omega^2\tau^2}\left(\frac{1}{E_\infty} + \frac{\omega^2\tau^2}{E_0}\right), \tag{1.81}$$

$$J'' = -\left(\frac{1}{E_\infty} - \frac{1}{E_0}\right)\frac{\omega\tau}{1+\omega^2\tau^2}. \tag{1.82}$$

Calculating $E^* = 1/J^*$, the moduli read

$$E' = E_\infty \frac{1 + (E_\infty/E_0)\omega^2\tau^2}{1 + (E_\infty/E_0)^2\omega^2\tau^2}, \tag{1.83}$$

$$E'' = E_\infty \frac{(1 - E_\infty/E_0)\omega\tau}{1 + (E_\infty/E_0)^2\omega^2\tau^2}. \tag{1.84}$$

Finally

$$\tan\delta = \frac{E''}{E} = \frac{(1 - E_\infty/E_0)\omega\tau}{1 + (E_\infty/E_0)\omega^2\tau^2}. \tag{1.85}$$

For $\omega\tau \to 0$, i.e. for very low frequencies and very long times, $E' \to E_\infty$ and $E'' \to 0$; for $\omega\tau \to \infty$, i.e. for very high frequencies and very short times, $E' \to E_0$ and $E'' \to 0$.

1.8.3 Hyperbolic Semiempirical Model

Cappella [58] has developed a semiempirical theory for the description of deformation–force curves on viscoelastic materials, which is exposed in the following with a new interpretation.

In this model the deformation D to the power of 3/2 as a function of the cantilever deflection δ ($D^{3/2}$-curve) is expressed for viscoelastic materials with a hyperbola in the form

$$\begin{aligned} D^{3/2} &= \beta\delta - \varepsilon + \sqrt{\alpha^2\delta^2 - 2\varepsilon(\beta-\gamma)\delta + \varepsilon^2} \\ &= \beta\delta - \varepsilon + \alpha\sqrt{(\delta - \delta_m)^2 + \frac{\varepsilon^2}{\alpha^2} - \delta_m^2}, \end{aligned} \tag{1.86}$$

with $\delta_m = \varepsilon(\beta - \gamma)/\alpha^2$.

Since the square root in Eq. (1.86) must yield real values of the deformation, the radicand must be positive for each value of δ, and this implies $\beta - \alpha < \gamma < \beta + \alpha$ or $-\varepsilon/\alpha < \delta_m < \varepsilon/\alpha$.

The first derivative of the $D^{3/2}$-curve is a monotonically increasing sigmoid with values between $\beta - \alpha$ and $\beta + \alpha$:

$$\frac{\partial D^{3/2}}{\partial \delta} = \beta + \frac{\alpha^2 \delta - \varepsilon(\beta - \gamma)}{\sqrt{\alpha^2 \delta^2 - 2\varepsilon(\beta - \gamma)\delta + \varepsilon^2}}$$
$$= \beta + \alpha \frac{\delta - \delta_m}{\sqrt{(\delta - \delta_m)^2 + \frac{\varepsilon^2}{\alpha^2} - \delta_m^2}}. \tag{1.87}$$

At $\delta = \delta_m$, $\partial D^{3/2}/\partial \delta = \beta$. Equation (1.87) shows that $\partial D^{3/2}/\partial \delta|_0 = \gamma$ and $\lim_{\delta \to \pm\infty} \partial D^{3/2}/\partial \delta = \beta \pm \alpha$.

Hence, the tangent of the curve at $\delta = 0$ is $D^{3/2} = \gamma\delta$ and the asymptotes of the hyperbola are

$$r_\pm = \left(\lim_{\delta \to \pm\infty} \frac{\partial D^{3/2}}{\partial \delta} \right) \delta + \lim_{\delta \to \pm\infty} \left(D^{3/2} - \delta \frac{\partial D^{3/2}}{\partial \delta} \right) = (\beta \pm \alpha)\delta - \varepsilon \mp \alpha \delta_m, \tag{1.88}$$

The asymptotes intersect in the point (δ_m, $\beta\delta_m - \varepsilon$), the tangent at $\delta = 0$ and the asymptote for $\delta \to \infty$ intersect at the point (ε/α, $\gamma\varepsilon/\alpha$).

Comparing Eq. (1.87) with Hertz theory $\partial D^{3/2}/\partial \delta = k_c/\left(E_{\text{tot}}\sqrt{R}\right)$, it is clear that, at a given temperature and frequency, the parameter γ is inversely proportional to the instantaneous elastic modulus $E_0(T, \nu)$ and the parameter $\beta + \alpha$ is inversely proportional to the equilibrium or infinite elastic modulus $E_\infty(T, \nu)$.

The second derivative is a bell function with a maximum corresponding to the flex point of the first derivative for $\delta = \delta_m$:

$$\frac{\partial^2 D^{3/2}}{\partial \delta^2} = \frac{\varepsilon^2 \left[\alpha^2 - (\beta - \gamma)^2\right]}{\sqrt{\alpha^2 \delta^2 - 2\varepsilon(\beta - \gamma)\delta + \varepsilon^2}^{\,3}} = \alpha \frac{\left(\frac{\varepsilon^2}{\alpha^2} - \delta_m^2\right)}{\left[(\delta - \delta_m)^2 + \frac{\varepsilon^2}{\alpha^2} - \delta_m^2\right]^{3/2}}. \tag{1.89}$$

The hyperbola and its first and second derivative are shown in Fig. 1.14.

The difference $\varepsilon^2/\alpha^2 - \delta_m^2$ is proportional to the width of the bell function and can be considered as the width of the transition region between instantaneous and equilibrium regime. If $\delta - \delta_m << -\sqrt{\varepsilon^2/\alpha^2 - \delta_m^2}$, the sample is elastic with modulus E_0. If $\delta - \delta_m \gg \sqrt{\varepsilon^2/\alpha^2 - \delta_m^2}$, the sample is elastic with modulus E_∞. In between, the modulus is time dependent and the $D^{3/2}$-curve shows a transition region.

The quantity $\varepsilon^2/\alpha^2 - \delta_m^2$, proportional to $\alpha^2 - (\beta - \gamma)^2$, shows whether the transition from E_0 to E_∞ occurs already at small forces ($\delta \to 0$) and short times.

When $\delta_m \to \pm\varepsilon/\alpha$, then $\gamma \to \beta \pm \alpha$, and the hyperbola in Eq. 1.86 tends to the composition of two straight lines with slopes $\beta \pm \alpha$ intersecting in $\delta_m = \pm\varepsilon/\alpha$, i.e. $D^{3/2} = [\beta + \alpha \text{sgn}(\delta - \delta_m)]\delta$. Also, the first derivative (Eq. 1.87) tends to a Heaviside step function $\partial D^{3/2}/\partial \delta = \beta + \alpha \text{sgn}(\delta - \delta_m)$ and the second derivative

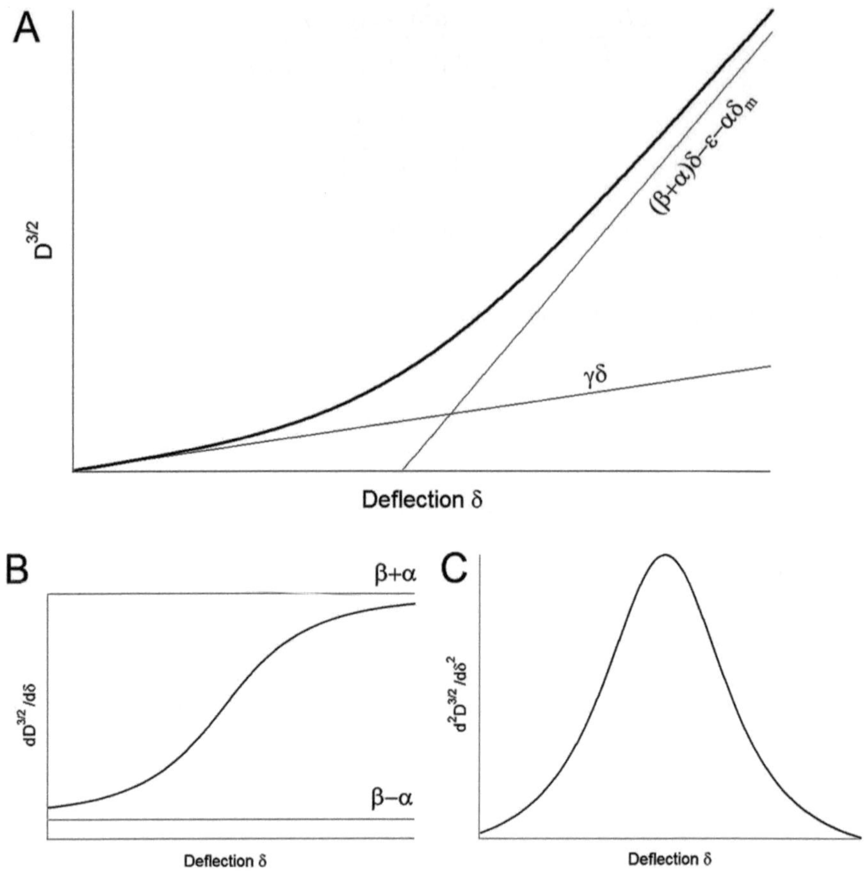

Fig. 1.14 The fitting hyperbola (Eq. 1.86) with the limiting lines for $\delta = 0$ and $\delta \to \infty$ (**A**) and its first (**B**) and second derivative (**C**)

(Eq. 1.89) to a Dirac function. In the first case ($\delta_m \to \varepsilon/\alpha$ and $\gamma \to \beta - \alpha$), the first derivative of the hyperbola goes from the value $\gamma \cong \beta - \alpha$ to the value $\beta + \alpha$ in a very short interval, centred at $\delta = \varepsilon/\alpha$. If the maximum attained deflection $\delta_{\max}$ is lower than ε/α, Eq. (1.86) is identical to Hertz equation, and the instantaneous elastic modulus E_0 is inversely proportional to the parameter $\beta - \alpha$.

In the second case ($\delta_m \to -\varepsilon/\alpha$ and $\gamma \to \beta + \alpha$), the transition region begins already at very small forces and short times and only E_∞, proportional to $\beta + \alpha$, can be measured.

In Fig. 1.15, five hyperbolas with different values of γ ($\gamma = \beta - 0.9\alpha$ (1), $\beta - 0.5\alpha$ (2), β (3), $\beta + 0.5\alpha$ (4) and $\beta + 0.9\alpha$ (5)) are shown together with the lines $D^{3/2} = (\beta \pm \alpha)\delta$ (dashed lines).

The parameter γ determines the position of the hyperbola and is an indicator of the mechanical behaviour of the sample. The curves with $\gamma \approx \beta \pm \alpha$, i.e. curves 1 and 5, are situated next to the lines $D^{3/2} = (\beta \pm \alpha)\delta$ and are fitted by the sections of

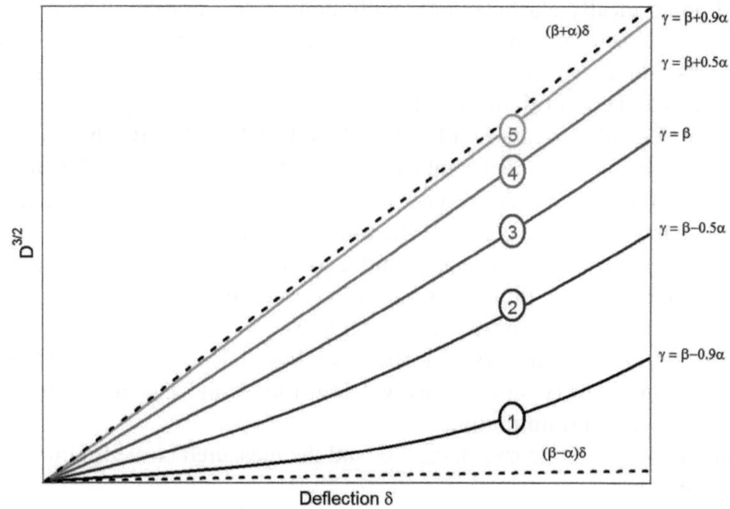

Fig. 1.15 Five hyperbolas with the ratio $(\gamma - \beta)/\alpha$ varying between −0.9 (1) and 0.9 (5), as shown in the legend. The lines $D^{3/2} = (\beta \pm \alpha)\delta$ are shown as *black dashed lines*

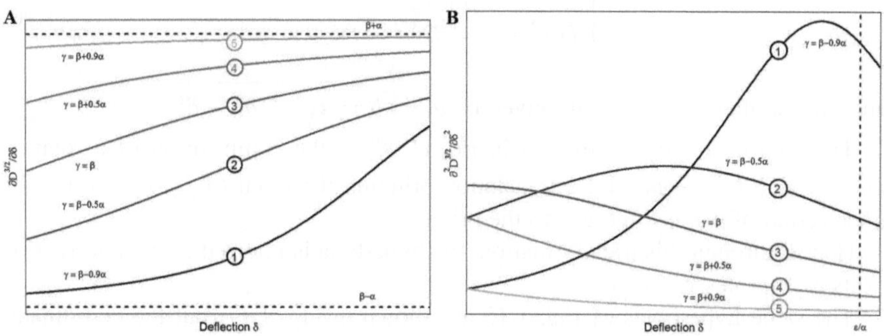

Fig. 1.16 (**A**) First derivative of the hyperbolas shown in Fig. 1.15, together with the instantaneous and equilibrium plateaus $\beta \pm \alpha$ (*black dashed lines*). (**B**) Second derivative of the hyperbolas shown in Fig. 1.15, together with the limit of the abscissa of the maximum, ε/α (*black dashed line*)

the hyperbola for $\delta \to \pm\infty$, which can be approximated with a line corresponding to an ideally elastic polymer with modulus E_0 or E_∞. Also, in curve 1, the transition can be seen clearly.

The role played by the parameter γ is confirmed by the first derivative of $D^{3/2}$, shown in Fig. 1.16A. Curve 5 has already at $\delta = 0$ a very large compliance, next to the maximum value $\beta + \alpha$, corresponding to E_∞. Hence, in curve 5, the transition to the equilibrium modulus occurs already at very short times and very small forces. On the contrary, curve 1 has at $\delta = 0$ a small compliance, next to the minimum value $\beta - \alpha$. Even if the compliance increases strongly for high loads and long times, at least the first half of the curve, before the transition starts, resembles a Hertz curve acquired on an ideally elastic sample with modulus E_0. For curve 3, at $\delta = 0$, the compliance is exactly intermediate between the minimum and maximum value. In other words, for curve 3, $\delta = 0$ is the centre of the transition region between instantaneous and equilibrium modulus.

Also the second derivatives of $D^{3/2}$, plotted in Fig. 1.16B, show quite well whether the polymer is in the instantaneous or in the infinite regime. The maximum of the second derivative is located at $\delta = \delta_m$, i.e. at the interception of the asymptotes, and is also the centre of the transition region.

When $\gamma \rightarrow \beta - \alpha$, as in curves 1 and 2 of Fig. 1.16, $\delta_m \rightarrow \varepsilon/\alpha$. Hence, the transition occurs at long times and high forces and at least the initial portion of the curve is situated in the instantaneous regime.

When $\gamma \rightarrow \beta + \alpha$, as in curves 4 and 5 of Fig. 1.16, $\delta_m \rightarrow -\varepsilon/\alpha$. Hence, the transition occurs at very short times and low forces and the measured curve is situated in the equilibrium regime.

Finally, if $\gamma \approx \beta$, as in curve 3, $\delta_m \approx 0$, and the measured curve is situated in the transition region.

The hyperbola in Eq. (1.86) can be put in the form $\frac{y^2}{a^2} - \frac{x^2}{b^2} = 1$ (or $y = \frac{a}{b}\sqrt{x^2 - a^2}$) through a coordinate transformation. Such a transformation is given by

$$\begin{cases} \delta \rightarrow x = \delta - \delta_m \\ D^{3/2} \rightarrow y = D^{3/2} - \beta\delta + \varepsilon \end{cases} \tag{1.90}$$

and the parameters a and b are given by $a = \alpha b = \alpha\sqrt{\varepsilon^2/\alpha^2 - \delta_m^2}$.

The coordinate transformation in Eq. (1.90) is the composition of a rotation $D^{3/2} \rightarrow Y = D^{3/2} - \beta\delta$ and a translation bringing the point $(\delta_m, \beta\delta_m + \varepsilon)$, i.e. the intersection of the asymptotes, to the origin.

Hence, through this transformation, the hyperbola is referred to the intersection of its asymptotes $y = \pm\alpha x$.

The same hyperbolas of Fig. 1.15 are shown in Fig. 1.17, but the coordinates have been transformed following Eq. (1.90). Again, the ratio $(\beta - \gamma)/\alpha$ is an indicator of the mechanical behaviour of the curves. When $\gamma \rightarrow \beta - \alpha$ and $\delta_m \rightarrow \varepsilon/\alpha$ (curve 1), the measured curve begins in the fourth quadrant, i.e. in the instantaneous regime. When $\gamma \rightarrow \beta + \alpha$ and $\delta_m \rightarrow -\varepsilon/\alpha$ (curves 4 and 5), the measured curve begins in the first quadrant, i.e. in the equilibrium regime. The other curves are situated in both quadrants or next to the Y-axis. Hence, they are in the transition region. The fundamental role of the parameter γ becomes even more evident in such a representation: it determines the position of the beginning of the curve and "shifts" the curves to the right, i.e. towards equilibrium, when going from $\beta - \alpha$ to $\beta + \alpha$.

In this model, the deformation is expressed as a function of the deflection or of the force and not of time. As already pointed out, deflection and force are not proportional to the time during an indentation. Hence, this model cannot be compared with the model based on a SLS unit.

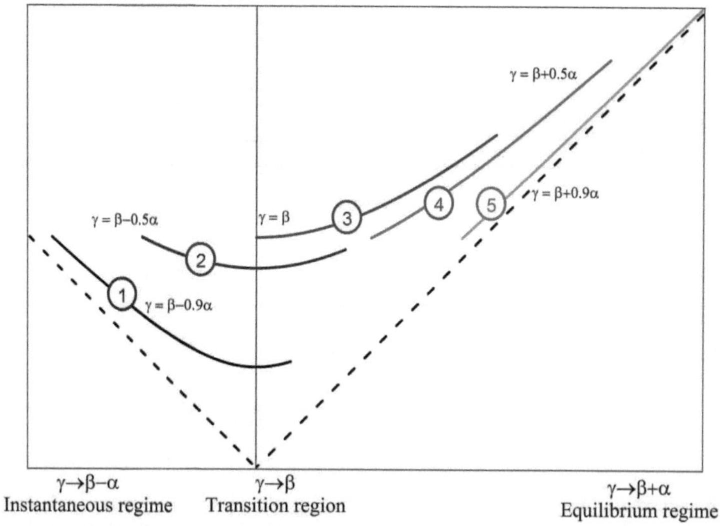

Fig. 1.17 The hyperbolas shown in Fig. 1.15 have been rotated and shifted so that they are referred to the intersection of their asymptotes $y = \pm \alpha x$, shown as *black dashed lines*. The curves go through the three mechanical regimes: the instantaneous regime (curve 1), the transition region (curves 2 and 3) and the equilibrium regime (curves 4 and 5)

1.8.4 Creep Compliance Measurement

A creep compliance measurement is designed for the analysis of the response of a polymer with viscoelastic behaviour to a step load. To this aim, a step load is applied to the Z-piezo, and a feedback loop is employed to maintain constant the cantilever deflection and hence the force applied. While keeping the force constant, the piezo elongation is detected as a function of time.

When consisting in a single-step load, a creep compliance measurement is actually the acquisition of a force distance curve with a so-called dwell time, which is a common option in current commercial microscopes. Yet, the "product" of such a measurement, used for the analysis, is not a force–distance curve, but a deformation–time curve. Nevertheless, since creep compliance measurements performed with AFM have been widely used to study the viscoelastic behaviour of polymer samples, some general features of this technique and two hands-on examples are included in this book (Chap. 6). More detailed information can be found in the review of Cohen and Kalfon–Cohen [59, 60, 61].

If the sample is described by an SLS model, the response to a step stress and the creep compliance are given by Eq. (1.75). Lu et al. [60] have computed the creep function also for measurements performed with a conical tip. Furthermore, they have pointed out that an ideal step load cannot be achieved in real measurements and have calculated the corrections needed to take into account the fact that the beginning and the end of the loading step are actually ramps with very short rise times.

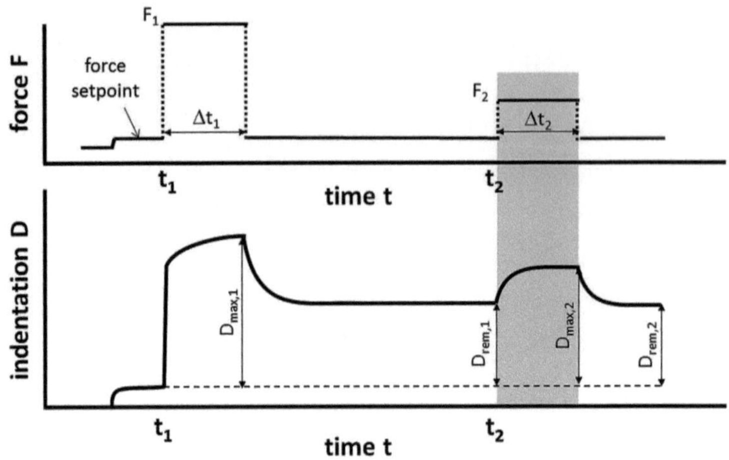

Fig. 1.18 Schematic description of the experimental protocol with two successive step loads at times t_1 and t_2. The *upper graph* shows the force and the lower graph the indentation or deformation, both as functions of time

Usually, the experimental protocol in a creep compliance measurement envisages two successive loading steps. The reason for this is that AFM indentations with sharp tips, most of all with conical tips, are likely to produce plastic deformations, which cannot be separated from the viscoelastic response, thus leading to erroneous interpretations of data.

As illustrated in the upper graph of Fig. 1.18, after establishing contact at a (small) setpoint force F_0, at the time t_1, a preload step with force F_1 is applied. After a time Δt_1, the force is reduced again to the setpoint value F_0. At the time t_2, the main step load with force F_2 and duration Δt_2 is applied.

The corresponding deformation or indentation, calculated as usually as $D = Z - \delta$, is shown in the lower graph of Fig. 1.18. During the preload step, the deformation grows to $D_{max,1}$. When the force is again brought to the setpoint value, the deformation partially relaxes. The remaining deformation $D_{rem,1}$ is due to irreversible plastic processes and to viscoelastic processes relaxing in a time longer than the time interval between the two step loads [62]. The preload step ensures that the additional deformation during the main step, $D_{max,2} - D_{rem,1}$, is due to viscoelastic processes and is not a plastic deformation.

1.8.5 Force Modulation

Like creep compliance measurements, force modulation measurements are, strictly speaking, not conventional force–distance measurements. Yet, this technique is aimed to the measurement of the complex modulus E^*, and as illustrated below, measurements are performed with a modification of the protocol used for the acquisition of force–distance curves. Hence, this technique is briefly discussed in the present section. More detailed information can be found in the review of Cohen and Kalfon–Cohen [59] and in references therein.

In a force modulation measurement, the sample is indented through a conventional force–distance curve with dwell time. During the dwell time, the force exerted by the cantilever is modulated at a given amplitude and frequency. The amplitude of the force and of the deformation as well as the phase shift of the cantilever response is monitored. Via a suitable rheological model for the description of the system, the complex elastic modulus can be calculated from the acquired signals.

It is important to remember that rheological models for AFM are substantially different from those for instrumented nanoindentation [63, 64], since in an AFM the indenting tip is fixed onto a deformable cantilever and is not a rigid punch. Hence, with a nanoindenter, the displacements of the tip and of the sample (i.e. the deformation) are always the same; with an AFM, the displacements of the tip and of the sample (i.e. deflection and deformation) differ, whereas the applied load is always the same.

As a consequence, an AFM system must be described through two Voigt–Kelvin elements in series: the first, with spring constant k_c and damping coefficient γ_c, represents the cantilever, and the second, with spring constant k_s and damping coefficient γ_s, represents the sample. The modulation load can be applied on the sample or on the cantilever. Rheological models and equations for both setups are described in [65].

The differential equation describing the system when the modulation is applied to the cantilever is

$$-m\ddot{D} + \gamma_c(\dot{Z} - \dot{D}) + k_c(Z - D) = \gamma_s\dot{D} + k_sD, \tag{1.91}$$

where $Z(t)$ and $D(t)$ are the displacement of the piezo and of the tip, i.e. the deformation, and m the mass of the cantilever.

If $Z(t)$ is a periodic function,

$$Z(t) = Z_0 + Z_1\exp(i\omega t), \tag{1.92}$$

with ω the angular frequency, the steady state function for the deformation D reads

$$D(t) = D_0 + D_1\exp[i(\omega t + \varphi)], \tag{1.93}$$

with φ the phase shift.

The time-independent deformation is

$$D_0 = \frac{k_c}{k_c + k_s}Z_0. \tag{1.94}$$

The time-dependent part can be described through following expressions for the ratio D_1/Z_1 and $\tan\varphi$:

$$\frac{D_1}{Z_1} = \frac{\sqrt{1 + \left(\frac{\gamma_c \omega}{k_c}\right)^2}}{\sqrt{\left(1 + \frac{k_s}{k_c}\frac{\omega^2}{\omega_0^2}\right)^2 + \left[\frac{\omega(\gamma_s + \gamma_c)}{k_c}\right]^2}}, \tag{1.95}$$

$$\tan\varphi = \frac{\omega\left[\gamma_c\left(k_c\frac{\omega^2}{\omega_0^2} - k_s\right) + k_c\gamma_s\right]}{k_c\left(k_c + k_s - k_c\frac{\omega^2}{\omega_0^2}\right) + \gamma_c\omega^2(\gamma_c + \gamma_s)}, \tag{1.96}$$

where $m = k_c/\omega_0^2$.

From these equations the elastic constant k_s and the damping coefficient γ_s of the sample can be calculated:

$$k_s = \frac{Z_1}{D_1}k_c\left(\cos\varphi + \frac{\omega\gamma_c}{k_c}\sin\varphi\right) - k_c + k_c\frac{\omega^2}{\omega_0^2}, \tag{1.97}$$

$$\gamma_s = \gamma_c\left[\frac{Z_1}{D_1}\left(\cos\varphi - \frac{k_c}{\omega\gamma_c}\sin\varphi\right) - 1\right]. \tag{1.98}$$

Finally, k_s and γ_s can be employed to determine the complex reduced modulus $K^* = K' + iK''$:

$$k_s = \frac{3}{2}aK', \tag{1.99}$$

$$\omega\gamma_s = \frac{3}{2}aK'', \tag{1.100}$$

with a the contact radius determined from Hertz, DMT or JKR theory.

In order to determine K^*, not only k_c must be calibrated (see Sect. 2.2.1) but also γ_c and ω_0. This can be done by fitting the resonance peak of the free oscillating cantilever. Modelling the cantilever as a harmonic oscillator, the oscillation amplitude δ_0 in air is given by

$$\delta_0 = \delta_i\frac{Q\omega_0^2}{\sqrt{\omega_0^2\omega^2 + Q^2(\omega_0^2 - \omega^2)}}, \tag{1.101}$$

where δ_i is the amplitude of the exciting oscillation and Q is the quality factor. From the definition of the quality factor, γ_c can be determined as

$$\gamma_c = \frac{k_c}{\omega_0 Q}. \tag{1.102}$$

In some other works [66], the oscillating force is applied at the end of the cantilever, and the response of the cantilever is measured out of contact as well as in contact with the sample. When the cantilever is out of contact, i.e. free oscillating, the cantilever can be modelled by a single Voigt element; in contact, two Voigt elements are needed for describing the system, but in this experimental setup, they are in parallel. Hence

$$F(t) = k_c \delta_0 \exp(i\omega t + \varphi_0) = (k_c + k_s)\delta_s \exp(i\omega t + \varphi_s). \tag{1.103}$$

where δ_0 and δ_s are the cantilever deflection amplitudes out of and in contact and φ_0 and φ_s are the phase shifts between the driving input and the cantilever response out of and in contact.

Putting $\overline{\delta} = \delta_0/\delta_s$ and $\overline{\varphi} = \varphi_0 - \varphi_s$, the real and imaginary part of the complex stiffness are given by

$$k_s{}' = k_c\left(\overline{\delta}\cos\overline{\varphi} - 1\right), \tag{1.104}$$

$$k_s{}'' = k_c \sin\overline{\varphi}. \tag{1.105}$$

Although in this configuration the analysis results much easier than in the previous one, applying the force at the end of the cantilever requires profound changes of the experimental setup. For example, in [66], magnetically driven cantilevers were used. A drive signal was applied to the gold-coated back side of the cantilevers so that the electromagnetic field generated at their end was perpendicular to the magnetic field generated by a magnet attached to the tip holder, engendering oscillations at the end of the cantilever.

The theoretical models of force modulation outlined above have several limitations:

1. Modelling the sample with a Voigt element is quite simplistic.
2. The adhesion is accounted for only in the calculation of the contact radius, when DMT or JKR is used, and not as additional force in the differential equation. Furthermore, the adhesion force depends also on frequency.
3. Plastic deformations, which are likely to occur, are not accounted for.

Most works employing force modulation have been performed with a nanoindenter [67–73]. Few works have been performed with AFM. In most of them, the samples are biological materials [74, 75]. In the very few publications, where force modulation is employed for the characterization of polymer samples [66, 76, 77], the analysis is affected by a severe misinterpretation. In these works the contact radius to be inserted in Eqs. (1.99) and (1.100) for the calculation of K' and K'' is determined through a fit of the retraction curve with JKR theory (see Sect. 2.3.1). Since measurements are performed in air, such estimation of the contact radius is subject to large errors.

1.9 Plastic Deformation

As already said, an elastic deformation is recovered when the load is removed. In contrast, a plastic deformation is not recovered when the load is again zero and is irreversible. A plastic deformation always implies energy dissipation, since work must be done to change irreversibly the mutual position of atoms and molecules in a solid material.

Since materials handled in this book are polymers, which exhibit viscoelastic behaviour, and deformations are measured through force–distance curves, such definition of a plastic deformation needs some specifications.

Due to the viscoelastic behaviour, the definition of plastic deformation becomes someway unclear, because it depends on the relaxation time typical for the material. The same deformation may appear irreversible, i.e. plastic, or reversible, i.e. elastic, if it is monitored for a time shorter or longer than the relaxation time. Since relaxation times may exceed 24 h [78], this is a relevant question for assessing the feasibility of an experimental characterization of plastic deformations.

An additional factor becomes important when deformations are characterized by means of force–distance curves. In a force–distance curve, deformations can be monitored only as long as the tip is in contact with the sample. Other than in a creep experiment, in a force–distance curve the tip is typically detached from the sample some microseconds after reaching the zero force along the retraction contact line (the time elapsing between the zero force and the detachment being determined by the distance at which the jump-off-contact occurs). Even if, after detachment, the deformation is recovered, a force–distance curve cannot "see" it. This has two consequences:

1. In several experiments performed with force–distance curves, a plastic deformation is defined as a deformation, which is not recovered when the load is removed and not as an irreversible deformation. As a consequence, the deformation of a viscoelastic material, which is recovered after the detachment of the tip, is considered as a plastic deformation, since the relaxation cannot be monitored.
2. The characterization of plastic deformations with an AFM requires other measurements additionally to force–distance curves.

The most common way to characterize plastic deformation is to image the sample topography after a force–distance curve.

Figure 1.19 shows Tapping Mode topographies of cavities carved on poly (methyl methacrylate) (PMMA) by force–distance curves with different maximum forces F_{max} [79].

It is evident that the depth and the lateral dimensions of the cavities increase with increasing maximum force. Furthermore, the first cavity, obtained with $F_{max} = 1.2$ μN, has a circular cross section, whereas the other ones have a triangular cross section. This is due to the shape of the tip, which has a hemispherical apex and a pyramidal shaft.

More important, all cavities are surrounded by increasing amounts of extruded material, called pile-up material. The pile-up material can be used to characterize plastic deformations.

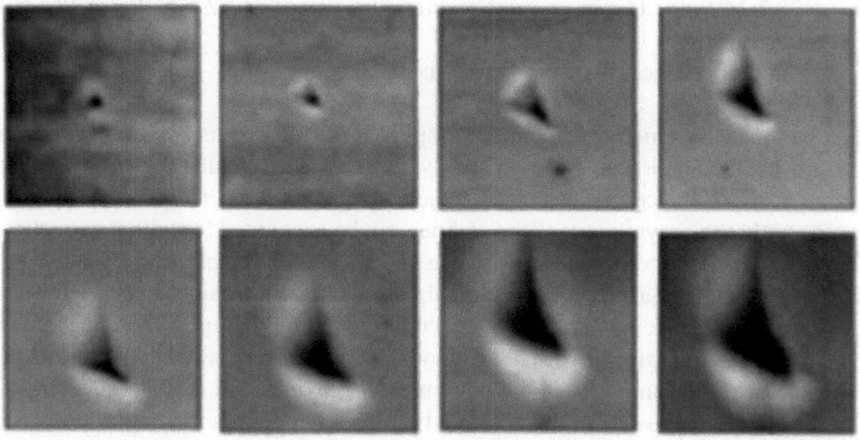

Fig. 1.19 Cavities carved on PMMA by force–distance curves with different maximum forces F_{max}. From *left* to *right*, *top row*, 1.2, 4, 8 and 12 μN; *bottom row*, 20, 24, 28 and 32 μN. Each Tapping Mode image is 700×700 nm^2 large. The *Z* scale of the topographies is 15 nm, 35 nm, 48 nm, 70 nm (from *left* to *right*, *top row*) and 160 nm (*bottom row*)

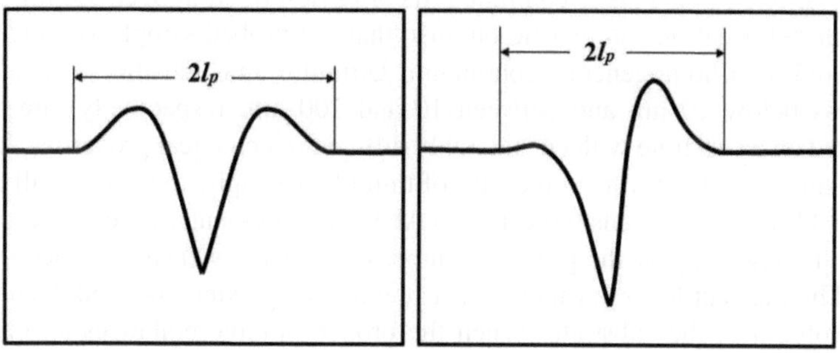

Fig. 1.20 Schematic representation of the plastic zone size for a symmetric (*left*) and an asymmetric (*right*) cavity

Du et al. [80] suggested defining the plastic zone size l_p by acquiring a topography line profile over the cavity. The plastic zone size $2l_p$ is the distance where the topography line profile of the sample surface differs from that of a flat surface, as illustrated schematically in Fig. 1.20 for a symmetric and an asymmetric cavity.

The plastic zone size l_p is related to the yield strength σ_y, i.e. the pressure, at which plastic deformations start occurring, by [81]

$$l_p = \sqrt{\frac{3F_{max}}{2\pi\sigma_y}}. \tag{1.106}$$

Du et al. have measured the yield strength of polycarbonate (PC) and polystyrene (PS) by means of the above equation, finding $\sigma_y = 141.2$ MPa for PC and

$\sigma_y = 178.7$ MPa for PS, in contrast to literature values of 72.5 MPa for PC and 88.5 MPa for PS.

The authors suggest two causes for this discrepancy:

1. Due to the asymmetric shape of the employed tip, all indents are asymmetric. In this case the applied pressure is not only directed perpendicular to the surface but has also a lateral component, against the assumptions of Eq. (1.106). The authors found that the asymmetry of the cavities can be accounted for with a correction factor depending only on the tip shape and not on the sample.
2. The surface is imaged with the same tip used to indent it. Since every deformation has an elastic component, the cavities carved by the tip, when imaged, are smaller than the tip. In particular, they have a narrower vertical cross section. Hence, the tip cannot reach the bottom of the cavities and the measurement of the lateral dimensions of the deformed sample is not correct.

1.10 Thin Polymer Films

The use of one of the elastic continuum theories described in Sect. 1.4 implies not only that deformations are elastic but also that the probed sample volume can be considered as a homogeneous continuum. Ultrathin and thin films (films with a thickness below 10 nm and between 10 and 100 nm, respectively) are usually deposited on a substrate with considerably different mechanical properties, forming a mechanical double layer. In the case of thin films, despite the very small contact areas and little indentations typical of an AFM measurement, the second condition, i.e. the homogeneity of the probed volume, is not always ensured, since the load applied by the cantilever during the measurement may extend beyond the interface and involve also the substrate. When the probed volume embraces an interface, elastic continuum theories cannot be employed for the description of sample deformations.

To solve this problem, either the probed volume must be reduced or a theory must be developed, which permits to extract the mechanical properties of the components (polymer and substrate) from the indentation curves acquired on a mechanical double layer.

The probed volume can be reduced by either decreasing the applied force or by increasing the contact area, i.e. by increasing the dimensions of the probe. Both approaches induce a decrease of the indentation depth. Yet, the more the indentation depth is decreased, the less information is obtained by the measurement. For example, in case of a 10 nm thick film, in order to avoid any influence of the substrate, the indentation depth should not exceed 1 nm. Due to the vertical resolution of an AFM, the measurement would yield data with hardly any significance.

In order to develop a theory predicting the deformation of a thin film deposited on a substrate, the sample is regarded as a system, whose mechanical properties are a composition of the properties of film and substrate. Moreover, the contributions of the components to the overall mechanical properties depend on the film thickness.

Several semiempirical approaches have been proposed in the past decades for the interpretation of the nanoindentation of compliant films on a stiff substrate.

All such equations express the deformation of the mechanical double layer or its modulus as a function of the film thickness t_f and of the elastic moduli of the film, in our case a polymer film, and of the substrate.

For all equations valid for mechanical double layers, when $t_f \to 0$, the deformation and the modulus must tend to the deformation that would be obtained on the substrate and to the elastic modulus of the substrate; in this case, the samples are "substrate dominated". Analogously, for $t_f \to \infty$, the deformation and the modulus must tend to the deformation that would be obtained on a homogeneous polymer sample and to the elastic modulus of the polymer; in this case, the samples are "polymer dominated".

For films of intermediate thickness, the deformation goes from the value that would be obtained on the homogeneous polymer to the value that would be obtained on the substrate depending on the ratio of the indentation to the film thickness or, in other words, depending on the load. At the beginning of the indentation, the tip probes only the polymer film, and deformation and stiffness are nearly the same as would be obtained on the polymer without substrate. By increasing the load, and hence the indentation, the tip probes more and more the underlying substrate. As a consequence, the stiffness increases, till, for very high loads, the same stiffness as on the substrate is measured. In other words, the elastic modulus as a function of the load shows a transition region between the value corresponding to the homogeneous polymer and the value corresponding to the substrate.

In the following the four most important models of the mechanical properties of mechanical double layers are listed, namely the equations of Tsukruk, Doerner and Nix, Gao, and Kovalev. Other similar functions [82–84] are discussed in [85].

Tsukruk Equation [86, 87]. Tsukruk equation expresses the ratio of the contact radius of a mechanical double layer, a, to Hertz contact radius $a_H = \sqrt[3]{RF/E_{\text{tot}}}$, i.e. the contact radius with a homogeneous polymer film without substrate, as a function of the reduced elastic moduli of polymer and substrate (E^p_{tot} and E^s_{tot}) and of the film thickness t_f:

$$\frac{a}{a_H} = \left[\frac{\left(E^p_{\text{tot}}/E^s_{\text{tot}}\right)^{4/3} + 0.8t_f/a_H}{\sqrt{1 + \left(0.8t_f/a_H\right)^2}}\right]^{1/4}. \tag{1.107}$$

Tsukruk equation cannot be employed for every kind of mechanical double layers. The equation was derived on the basis of measurements on samples consisting of materials with very different Young's moduli (E_p/E_s ranging from 0.1 to 0.01). The ratio a/a_H, regardless of the film thickness, should be always smaller than 1, also for $E_p \approx E_s$, because the substrate always has the effect of decreasing the deformation. This condition is not met by Eq. (1.107), since, for $E^p_{\text{tot}}/E^s_{\text{tot}} \to 1$, $\left[\left(E^p_{\text{tot}}/E^s_{\text{tot}}\right)^{4/3} + 0.8t_f/a_H\right]^2$ is greater than $1 + \left(0.8t_f/a_H\right)^2$.

Equation of Doerner and Nix [88]. In this model the reduced elastic modulus of a mechanical double layer is expressed as the superposition of the reduced elastic moduli of polymer and substrate weighted with the factor $\exp\left(-\alpha t_f/D\right)$:

$$\frac{1}{E_{\text{tot}}} = \frac{1}{E^p_{\text{tot}}}\left[1 - \exp\left(-\frac{\alpha t_f}{D}\right)\right] + \frac{1}{E^s_{\text{tot}}}\left[\exp\left(-\frac{\alpha t_f}{D}\right)\right]. \tag{1.108}$$

The parameter α accounts for the interactions at the film–substrate interface. The larger the adhesion between film and substrate, the better the load applied onto the film is "transmitted" to the substrate. The parameter α is specific for every material couple, but it must be determined experimentally and there is no theoretical approach to calculate it. Hence, α is an additional parameter in the fit and interpretation of the experimental data. As a consequence, the original task of such an equation, i.e. determining the film thickness when the moduli of the components are known or determining the elastic modulus of the polymer when the modulus of the substrate and the thickness are known, is not achieved.

Gao Equation [89]. In Gao equation, the reduced elastic modulus is expressed in the form

$$E_{\text{tot}} = E^p_{\text{tot}} + \left(E^p_{\text{tot}} - E^s_{\text{tot}}\right)\Phi(x), \tag{1.109}$$

with

$$\Phi(x) = \frac{2}{\pi}\arctan\left(\frac{1}{x}\right) + \frac{1-2\nu}{2\pi(1-\nu)}\ln\left(1+x^2\right) - \frac{x}{1+x^2}, \tag{1.110}$$

and $x = D/t$ or $x = a/t$.

This function has the advantage of depending only on one very well-known parameter, i.e. Poisson's ratio ν.

Kovalev Equation [87]. Other than Gao equation, this function depends on two parameters, λ and τ, which have no physical meaning and must be determined experimentally:

$$E_{\text{tot}} = E^p_{\text{tot}} + \frac{\left(E^p_{\text{tot}} - E^s_{\text{tot}}\right)}{1 + \exp\left(-\lambda \frac{E^s_{\text{tot}} - E^p_{\text{tot}}}{E^p_{\text{tot}}} \frac{x-\tau}{\tau}\right)}. \tag{1.111}$$

Again, it is $x = D/t$ or $x = a/t$.

These four equations have been tested in [90]. In order to summarize here the results, it would be necessary to anticipate some notions about the acquisition and the analysis of deformation curves, which are introduced in Chap. 2. Therefore, experimental curves are presented without details about sample preparation and data analysis. Since the same measurement is outlined in hands-on example 12, such details can be found in Sect. 4.2.

Figure 1.21 shows the deformation D (panel A) and the contact radius a (panel B) versus the cantilever deflection δ obtained on poly(n-butyl methacrylate) (PnBMA) films on glass. The thickness of the films is $t_f = 10$, 18, 26, 38, 50, 60, 67, 110, 190 and 430 nm. Also the curve on glass is shown.

For the 430 nm thick film, the thickness is so much larger than the maximum deformation (25 nm) that the sample can be considered as homogeneous, and the deformation curve obtained on it can be fitted with Hertz equation $D = \left(\frac{k_c \delta}{E_{\text{tot}} \sqrt{R}}\right)^{2/3}$ (Eq. 1.19). With $R = 25$ nm and $k_c = 37$ N/m, the fit, shown in Fig. 1.21 with a continuous line, yields $E_{\text{tot}} = 5.6 \pm 0.3$ GPa. The relation $\frac{1}{E_{tot}} = \frac{3}{4}\left(\frac{1-\nu^2}{E} + \frac{1-\nu_t^2}{E_t}\right)$ (Eq. 1.12) with $E_t = 245$ GPa, $\nu_t = 0.27$ and $\nu = 0.5$ yields for the elastic modulus of PnBMA the value $E = 3.14 \pm 0.2$ GPa.

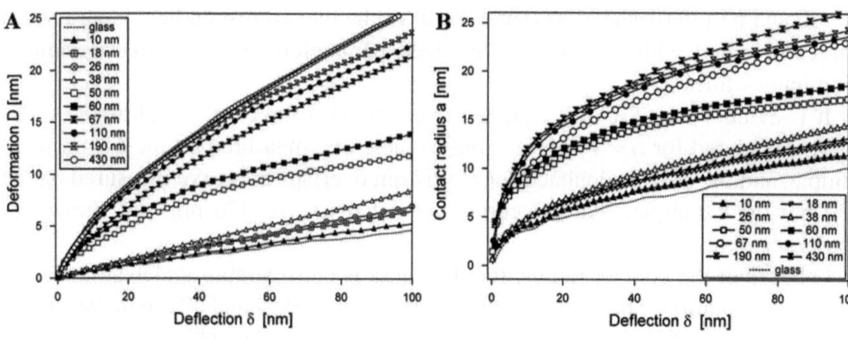

Fig. 1.21 (**A**) Deformation D versus deflection δ obtained on PnBMA films on glass with thickness $t_f = 10$, 18, 26, 38, 50, 60, 67, 110, 190 and 430 nm. Also the curve on glass is shown. The curve relative to the 430 nm thick sample is shown together with the fit (*continuous line*) with Hertz equation. (**B**) Contact radius a versus deflection δ obtained on the same samples. Also the reference curve on glass is shown. Reprinted with permission from [90]. Copyright 2008. Elsevier

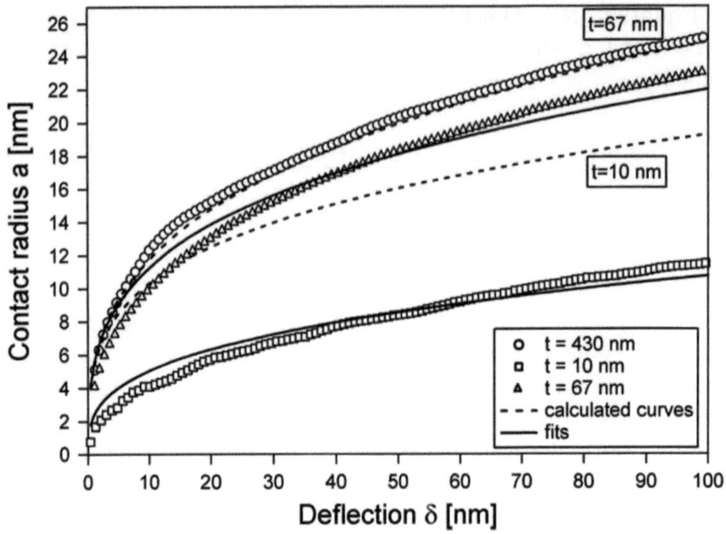

Fig. 1.22 Contact radii of the 10, 67 and 430 nm thick films plotted versus the deflection by *markers*. Curves calculated by means of Tsukruk equation for $t_f = 10$ nm and $t_f = 67$ nm are shown by *dashed lines*. The fits obtained by holding all parameters except the thickness are represented for $t_f = 10$ nm and $t_f = 67$ nm by *solid lines*. Reprinted with permission from [90]. Copyright 2008. Elsevier

Also the curve on glass was fitted with Hertz theory, yielding $E_{\mathrm{tot}} = 80$ GPa and $E = 72$ GPa.

The curves $a(\delta)$, calculated via the relation $a = \sqrt{DR}$, can be used to test the validity of Tsukruk equation. Since all fit parameters are known ($E^p_{tot} = 5.6$ GPa, $E^s_{\mathrm{tot}} = 80$ GPa and t_f), it is possible to calculate the radii predicted by Tsukruk equation with the known values. Figure 1.22 shows two of the calculated curves for $t_f = 10$ and 67 nm (dashed lines) and the fit of the two corresponding experimental curves obtained with E^p_{tot} and E^s_{tot} as fixed parameters and t_f as free parameter (continuous lines).

It is evident that Tsukruk equation overestimates the contact radius [87]. The curve calculated for $t_f = 10$ nm is close to the curve measured on the 67 nm thick sample, and the curve calculated for $t_f = 67$ nm overlaps the curve measured on the 430 nm thick sample, like the curve calculated for $t_f = 430$ nm, not shown for clarity.

Other than the contact radius, the thickness is considerably underestimated by Tsukruk's equation. The thickness yielded by the fit is 0.04 and 20 nm in the case of the 10 and 67 nm thick films. Also, the shape of both experimental curves cannot be reproduced by Tsukruk equation, predicting larger contact radii at small loads and smaller contact radii at large loads. By considering the fit of all curves shown in Fig. 1.21A, it becomes evident that the thickness predicted by Tsukruk equation is smaller than 30 % of the real thickness in all cases, even smaller than 3 % for $t_f < 40$ nm.

It can be concluded that the equation proposed by Tsukruk is not a reliable instrument either to predict the shape of experimental curves measured on mechanical double layers or to relate the film thickness t_f to the mechanical properties of a mechanical double layer.

In order to test the equations proposed by Doerner and Nix, Gao and Kovalev, the reduced elastic modulus E_{tot} has to be calculated from the experimental curves using the relation

$$E_{\text{tot}} = \frac{k_c}{\sqrt{R}} \frac{\partial \delta}{\partial D^{3/2}}. \tag{1.112}$$

Since $\partial \delta / \partial D^{3/2}$ is affected by rather large noise, the raw data have been smoothed prior to the calculation by averaging the deflection in regular intervals of the deformation.

Figure 1.23A shows the smoothed curves of the reduced elastic modulus E_{tot} versus the deformation D for selected films ($t_f = 18$, 50, 60, 67 and 430 nm). The value obtained on the 430 nm thick film is actually the value of the homogeneous PnBMA, $E^p_{\text{tot}} = 5.6$ GPa.

The dependence of the curves on the film thickness is as expected. The mechanical properties of the sample with a 67 nm thick polymer film, and also of all samples with thicker films, whose reduced elastic modulus is not shown for clarity, are governed by the PnBMA phase. In the case of the thinnest sample ($t_f = 18$ nm), the mechanical properties are totally governed by the substrate. The curves obtained on the 50 nm and 60 nm thick films show an intermediate behaviour: at

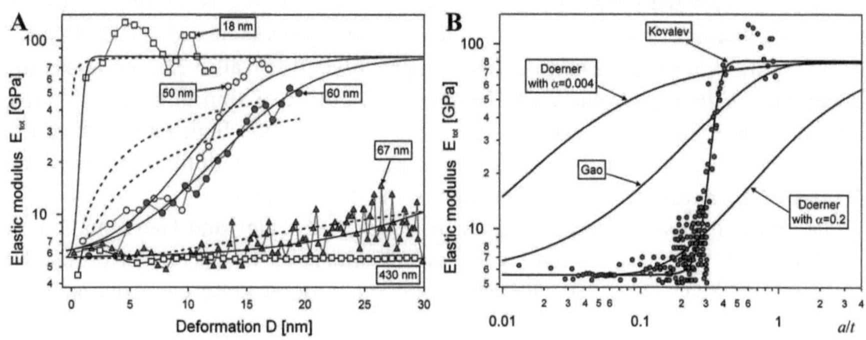

Fig. 1.23 (**A**) Reduced elastic modulus E_{tot}, obtained from curves on selected samples ($t_f = 18$, 50, 60, 67 and 430 nm), in logarithmic scale versus the deformation D (*markers*). The *dashed* and *solid lines* show the fits with Doerner and Kovalev equations, respectively. (**B**) Reduced elastic modulus E_{tot}, obtained from the same samples, in logarithmic scale versus the ratio a/t_f, yielding a kind of master curve of E_{tot}. The *solid lines* show the fits with Doerner equation (with two different values of the parameter α) and with Kovalev and Gao equations, as indicated. Reprinted with permission from [90]. Copyright 2008. Elsevier

low loads, the mechanical properties are dominated by the PnBMA phase, whereas, at higher forces, when the stress field extends to the substrate, E_{tot} is strongly affected by the substrate.

In Fig. 1.23A also the fits of the experimental curves both with Doerner (dashed lines) and Kovalev equation (dotted lines) are shown. Both fits were performed with the known values of E_{tot} for glass and for PnBMA. For Doerner equation one free parameter $\beta = \alpha t_{fit}$ is needed, whereas for Kovalev equation two free parameters λ and τ are required.

It is evident that Doerner equation can reproduce the experimental curves on very thin films (e.g. $t_f = 18$ nm) and on thick films (e.g. $t_f = 67$ nm). On the contrary, for $t_f = 50$ and 60 nm, the equation of Doerner cannot fit the experimental curves. In other words, Doerner equation can be used only to fit curves acquired on samples whose mechanical properties are governed by one of the two constituents, either the substrate or the polymer.

The fit with Doerner equation was performed for all curves obtained on the films with different thickness. If the fit were able to yield the correct values for the film thickness, it would be $t_{fit} = t_f$. Hence, it would be $\beta/t_f = \alpha$, and since all samples consist of the same material couple, β/t_f should be the same for all curves. On the contrary, α increases with the film thickness, from $\alpha \cong 10^{-3}$ for $t_f < 40$ nm up to $\alpha \cong 10^{-1}$ for $t_f > 70$ nm. As a consequence, if all experimental curves are fitted with the same averaged value of α, the film thickness of curves dominated by the substrate (polymer) would be severely underestimated (overestimated) by the fit.

Considering now the fit with Kovalev equation, it can be noted that this function is able to reproduce the shape of the curves much better than Doerner equation. Yet, a good fit is made possible by the fact that the equation contains two parameters λ and τ, which can be arbitrarily varied, since they are not related to the thickness or to other physical quantities characterizing the sample.

In Fig. 1.23B the reduced elastic modulus of all samples is plotted versus the ratio a/t_f, yielding a kind of master curve. The curves acquired on thick films ($t_f > 70$ nm), whose reduced modulus is approximately equal to E^p_{tot}, compose the lower part of the master curve ($0.01 < a/t_f < 0.2$), those on thin films ($t_f < 40$ nm), with a reduced modulus nearly equal to E^s_{tot}, make up the high ratio part of the master curve ($0.5 < a/t_f < 1$), and curves acquired on films with thickness between 40 and 70 nm form the intermediate part of the master curve.

The master curve has been fitted with Doerner, Kovalev and Gao equation. A similar fit of the master curve with several equations has been performed by Clifford and Seah [85], yet with data obtained through finite elements analysis methods.

As already pointed out, Kovalev equation fits the curve in the whole range of the ratio a/t_f, whereas Doerner equation fits either the low portion of the master curve (with $\alpha = 0.2$) or the high portion (with $\alpha = 0.004$).

Gao function is not able to fit the whole experimental master curve. Since this function has no free parameter, it can fit only one short portion of the curve, namely, the higher ratio portion.

Summarizing, it can be assessed that:

1. Tsukruk equation cannot be used as a global model, since it fails for higher ratios of elastic moduli of film and substrate. More important, this equation underestimates the thickness of the film and overestimates the contact radius.
2. If the function contains no free fitting parameters, such as Gao function, only curves acquired in one limiting case, i.e. thin or thick samples, can be fitted satisfyingly, if at all.
3. Doerner equation can fit both limiting cases, but not the intermediate region. Also, the fit of both limiting regimes must be performed with very different values of the fitting parameter α, despite the fact that this parameter, due to its physical meaning, should not change when the components of the double layer are the same.
4. Kovalev function is able to fit Young's modulus curves acquired on samples with very different thicknesses. Yet, this is due to the fact that the equation contains free parameters without any relation to physical quantities characterizing the sample.

Cappella has proposed a novel semiempirical function [91, 92] for the description of the deformation of a mechanical double layer. Even if the model is very similar to the model of the deformations of a viscoelastic material exposed in Sect. 1.8.3, since the physical meaning of the equations is completely different, it is exposed again in detail in the following.

In this model, the load dependence of the deformation of a mechanical double layer is described with a hyperbola in the form

$$\begin{aligned} D^{3/2} &= \beta\delta + \varepsilon - \sqrt{\alpha^2\delta^2 + 2\varepsilon(\beta-\gamma)\delta + \varepsilon^2} \\ &= \beta\delta + \varepsilon - \alpha\sqrt{(\delta-\delta_m)^2 + \frac{\varepsilon^2}{\alpha^2} - \delta_m^2}, \end{aligned} \quad (1.113)$$

where D is the deformation, $\delta = F/k_c$ the cantilever deflection and the parameter δ_m is given by $\delta_m = -\varepsilon(\beta-\gamma)/\alpha^2$. The hyperbolic function is shown in Fig. 1.24A.

The first derivative $\partial D^{3/2}/\partial\delta$, shown in Fig. 1.24B, is given by

$$\begin{aligned} \frac{\partial D^{3/2}}{\partial\delta} &= \beta - \frac{\alpha^2\delta + \varepsilon(\beta-\gamma)}{\sqrt{\alpha^2\delta^2 + 2\varepsilon(\beta-\gamma)\delta + \varepsilon^2}} \\ &= \beta - \alpha\frac{\delta-\delta_m}{\sqrt{(\delta-\delta_m)^2 + \frac{\varepsilon^2}{\alpha^2} - \delta_m^2}}. \end{aligned} \quad (1.114)$$

Hence, the first derivative is a monotonically decreasing sigmoid function with plateaus $\beta \mp \alpha$ for $\delta \to \pm\infty$. The plateau values are the slopes of the two asymptotes of the hyperbola, given by

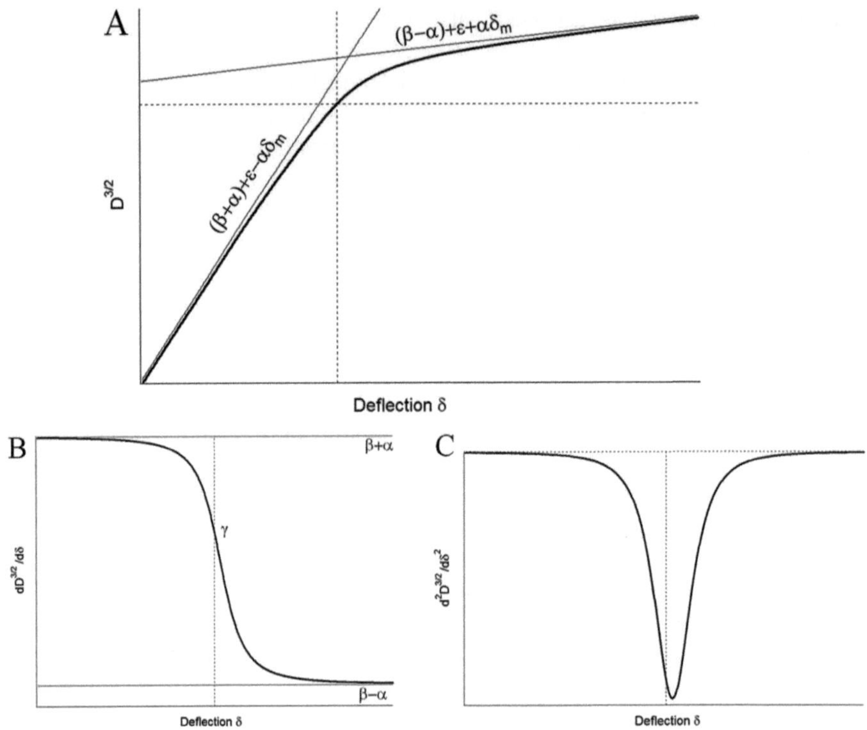

Fig. 1.24 The fit hyperbola with its asymptotes (**A**), the first (**B**) and the second derivative (**C**)

$$r_{\pm} \equiv D^{3/2} = (\beta \mp \alpha)\delta + \varepsilon\left(1 \mp \frac{\beta - \gamma}{\alpha}\right) = (\beta \mp \alpha)\delta + \varepsilon \pm \alpha\delta_m. \tag{1.115}$$

The intersection of the two asymptotes is the point with $\delta = \delta_m$ and $D^{3/2} = \beta\delta_m + \varepsilon$.

The slopes of the two asymptotes must be inversely proportional to the elastic moduli of substrate and polymer:

$$\beta - \alpha = \frac{k_c}{\sqrt{R}E^s_{\text{tot}}}, \quad \beta + \alpha = \frac{k_c}{\sqrt{R}E^p_{\text{tot}}}. \tag{1.116}$$

As a matter of fact, the two asymptotes are the translated $D^{3/2}$ curves corresponding to the substrate and to the homogeneous polymer.

The two parameters β and α are known, once Young's moduli of polymer and substrate are known.

The second derivative of the fit hyperbola, shown in Fig. 1.24C, is given by

$$\frac{\partial^2 D^{3/2}}{\partial \delta^2} = -\frac{\varepsilon^2\left[\alpha^2 - (\beta-\gamma)^2\right]}{\sqrt{\alpha^2\delta^2 + 2\varepsilon(\beta-\gamma)\delta + \varepsilon^2}^3} = -\alpha\frac{\frac{\varepsilon^2}{\alpha^2} - \delta_m^2}{\sqrt{(\delta-\delta_m)^2 + \frac{\varepsilon^2}{\alpha^2} - \delta_m^2}^3}. \quad (1.117)$$

This is a negative bell function, tending to 0 for $\delta \to \pm\infty$. It has a minimum $\partial D^{3/2}/\partial\delta = -\alpha/\sqrt{\varepsilon^2/\alpha^2 - \delta_m^2}$ at the intersection of the asymptotes, i.e. for $\delta = \delta_m$. The difference $\varepsilon^2/\alpha^2 - \delta_m^2$ is proportional to the width of the bell function and can be regarded as the width of the transition region between polymer-dominated and substrate-dominated regime, i.e. the width of the transition region between the portion of the curve next to the polymer asymptote and the one next to the substrate asymptote. When $\delta - \delta_m << -\sqrt{\varepsilon^2/\alpha^2 - \delta_m^2}$, the tip senses only the polymer film. When $\delta - \delta_m \gg \sqrt{\varepsilon^2/\alpha^2 - \delta_m^2}$, the tip senses only the substrate. In between, the $D^{3/2}$ curve shows a transition region.

The parameter γ (or δ_m) has a central meaning for the shape of the function. The parameter γ is the value of the first derivative for $\delta = 0$. Since the radicand in Eq. (1.113) must be positive, it must be $\beta - \alpha < \gamma < \beta + \alpha$ (or $-\varepsilon/\alpha < \delta_m < \varepsilon/\alpha$). This mathematical condition has a physical meaning: the elastic modulus of the sample probed at very small deformations must be intermediate between that of the homogeneous polymer and that of the substrate.

Both in the formula of Tsukruk and in Doerner equation is $E = E_p$ for $D = 0$ or $\delta = 0$. From a mathematical point of view, an infinitely small deformation is possible. Yet, from a physical point of view, an infinitely small deformation, i.e. a deformation being much smaller than the film thickness t_f, may be such that it cannot be measured or it cannot be described by elastic continuum theories, i.e. theories that do not account for the discrete composition of the sample. This is the case for ultrathin films ($t_f < 10$ nm). This means that the measured mechanical properties of an ultrathin film are always influenced by the substrate.

Turning back to the meaning of γ, there are two important limits of this parameter.

If $\gamma = \beta - \alpha$ (or $\delta_m = -\varepsilon/\alpha$), the hyperbola in Eq. (1.113) degenerates into the line $D^{3/2} = (\beta - \alpha)\delta$ and coincides with the Hertz straight line corresponding to the substrate. This means that the sample is only the substrate or, in other words, $t_f = 0$. If $\gamma = \beta + \alpha$ (or $\delta_m = \varepsilon/\alpha$), the hyperbola degenerates into the line $D^{3/2} = (\beta + \alpha)\delta$. This means that the sample is only the polymer or, in other words, $t_f \to \infty$. Yet, in this case, it must be $\delta < \varepsilon/\alpha$, so that the radicand in Eq. (1.113) is positive. This shows that also the parameter ε increases with increasing thickness and must tend to infinity when γ tends to $\beta + \alpha$.

In Fig. 1.25, five hyperbolas with different values of γ ($\gamma = \beta - 0.9\alpha$ (1), $\beta - 0.5\alpha$ (2), β (3), $\beta + 0.5\alpha$ (4) and $\beta + 0.9\alpha$ (5)) are shown. Also, the polymer and substrate asymptotes are shown with dashed lines.

It is evident that γ determines the position of the hyperbola and shows the mechanical behaviour of the sample. The curves with $\gamma \approx \beta \pm \alpha$, i.e. curves 1 and

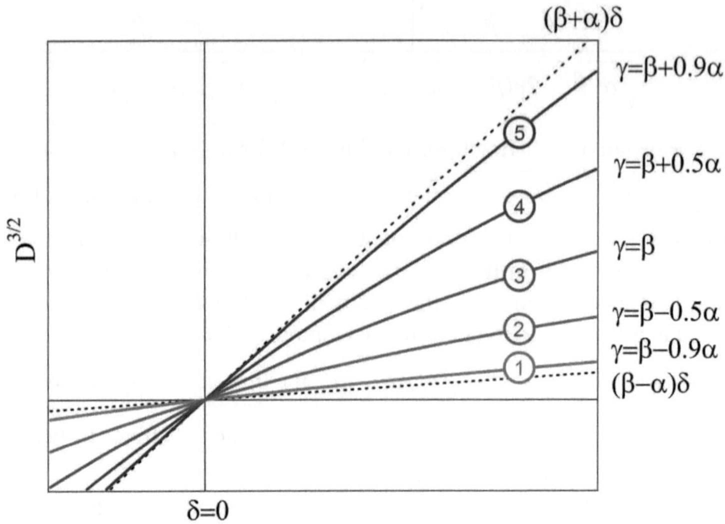

Fig. 1.25 Five hyperbolas with the ratio $(\beta - \gamma)/\alpha$ varying between −0.9 (1) and 0.9 (5), as shown in the legend. Also Hertz curves of substrate and bulk polymer, $D^{3/2} = (\beta \pm \alpha)\delta$, are shown (*black dashed lines*)

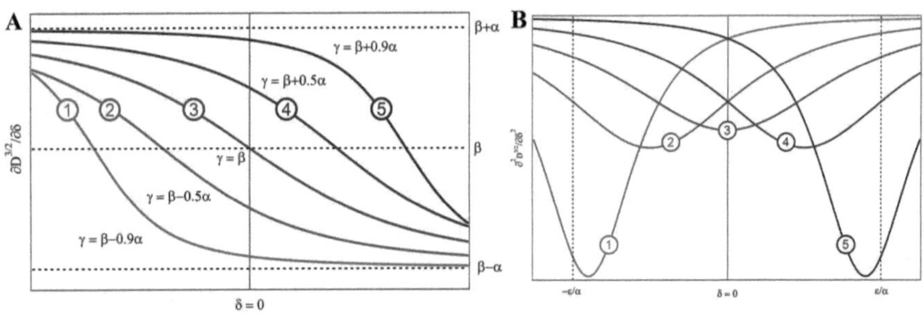

Fig. 1.26 (**A**) First derivative of the hyperbolas shown in Fig. 1.25, together with the polymer and substrate plateaus $\beta \pm \alpha$ (*black dashed lines*). (**B**) Second derivative of the hyperbolas shown in Fig. 1.25, together with the limits of the abscissa of the minimum, $\pm\varepsilon/\alpha$ (*black dashed lines*)

5, are situated next to the Hertz curves $D^{3/2} = (\beta \pm \alpha)\delta$ and are fitted by the sections of the hyperbola for $\delta \to \pm\infty$, which can be approximated with the polymer or substrate asymptote. The curve with $\gamma = \beta$, (curve 3), shows on the other hand a clear mixed behaviour.

The role played by the parameter γ is confirmed by the first derivative of $D^{3/2}$, shown in Fig. 1.26A. The first derivative $\partial D^{3/2}/\partial\delta$ is inversely proportional to Young's modulus and proportional to the compliance. At $\delta = 0$, curve 1 has already a very small compliance, next to the minimum value $\beta - \alpha$. The compliance can only decrease with increasing load; hence, curve 1 is substrate dominated.

On the contrary, curve 5 has at $\delta = 0$ a large compliance, next to the maximum value $\beta + \alpha$. Even if the compliance decreases strongly for high loads, this curve is polymer dominated. For curve 3, at $\delta = 0$, the compliance is exactly intermediate between the minimum and maximum value. In other words, for curve 3, $\delta = 0$ is the centre of the transition region between polymer-dominated and substrate-dominated regime.

This can be seen still better by considering the second derivative of $D^{3/2}$, shown in Fig. 1.26B. The minimum of the second derivative is located at $\delta = \delta_m$, i.e. at the interception of the asymptotes, and is also the centre of the transition region.

When $\gamma \to \beta - \alpha$, as in curves 1 and 2 of Fig. 1.26, $\delta_m \to -\varepsilon/\alpha$. Hence, the transition region is situated in the third quadrant ($\delta < 0$), and the measured curve, for which is $\delta > 0$, is situated in the portion of the hyperbola dominated by the substrate.

When $\gamma \to \beta + \alpha$, as in curves 4 and 5 of Fig. 1.26, $\delta_m \to +\varepsilon/\alpha$. Hence, the transition region is situated in the first quadrant ($\delta > 0$), and the measured curve is situated in the portion of the hyperbola dominated by the polymer. Finally, if $\gamma \approx \beta$, as in curve 3, $\delta_m \approx 0$ and the measured curve is situated in the transition region.

The hyperbola in Eq. (1.113) can be put in the form $y = -\frac{a}{b}\sqrt{x^2 - b^2}$ (or $\frac{x^2}{a^2} - \frac{y^2}{b^2} = 1$) through a coordinate transformation:

$$\begin{cases} D^{3/2} \to y = D^{3/2} - \beta\delta - \varepsilon \\ \delta \to x = \delta - \delta_m \end{cases} \tag{1.118}$$

The parameters a and b are given by $a = \alpha b = \alpha\sqrt{\varepsilon^2/\alpha^2 - \delta_m^2}$ and the hyperbola is written as

$$y = -\alpha\sqrt{x^2 - \frac{\varepsilon^2}{\alpha^2} + \delta_m^2} = -\alpha\sqrt{x^2 + \frac{\varepsilon^2}{\alpha^4}\left[(\beta - \gamma)^2 - \alpha^2\right]}. \tag{1.119}$$

The coordinate transformation in Eq. (1.118) is the composition of a rotation $D^{3/2} \to Y = D^{3/2} - \beta\delta$ and a translation bringing the point with $\delta = \delta_m$ and $D^{3/2} = \beta\delta_m + \varepsilon$, i.e. the intersection of the asymptotes, to the origin.

Hence, through this transformation, the hyperbola is referred to the intersection of its asymptotes $y = \pm\alpha x$, which coincides with the origin. Curves acquired on double layers with the same constituents, but different thickness, have the same asymptotes and can be perspicuously compared with each other in such a representation.

The same hyperbolas of Fig. 1.25 are shown in Fig. 1.27, but the coordinates have been transformed following Eq. (1.118). The thin black curves represent the fit hyperbola, whereas the thick curves in different grey scale represent the measured curves in the range $0 < \delta < \delta_{max}$. Since the equation of the hyperbola in this form depends only on the square of the difference $\beta - \gamma$, the curves 1 and 5 as well as the

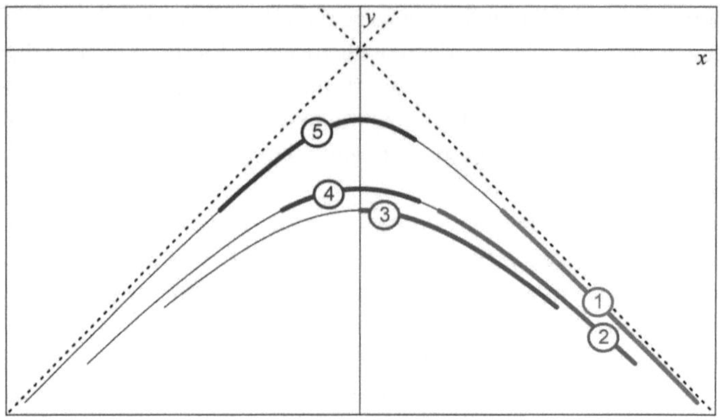

Fig. 1.27 The hyperbolas shown in Fig. 1.25 have been rotated and shifted so that they are referred to the intersection of their asymptotes $y = \pm\alpha x$, shown as *black dashed lines*. The *thin black curves* are the whole fitting hyperbolas, whereas thick curves in different grey scale are the measured curves in the range $(0, \delta_{\max})$. Since the equation of the hyperbola in this form depends only on the square of the difference $\beta - \gamma$, the curves 1 and 5 and 2 and 4 coincide. The measured curves go through the three mechanical regimes: the substrate-dominated regime (curves 1 and 2), the regime of mixed properties (curves 3 and 4) and the polymer-dominated regime (curve 5)

curves 2 and 4 lie on the same fit hyperbola. In other words, the deformations of two double layers with the same constituents but with different film thickness are described by the same hyperbola referred to its axis; the curves differ only in the beginning of the measured range, whose abscissa is $\delta = \delta_m$. Again, the ratio $(\beta - \gamma)/\alpha$ is an indicator of the mechanical behaviour of the curves. When $\gamma \to \beta + \alpha$ and $\delta_m \to -\varepsilon/\alpha$ (curve 5), the measured curve begins in the third quadrant, i.e. in the polymer-dominated quadrant. When $\gamma \to \beta - \alpha$ and $\delta_m \to +\varepsilon/\alpha$ (curves 1 and 2), the measured curve begins in the fourth quadrant, i.e. in the substrate-dominated quadrant. The other curves are situated in both quadrants or next to the Y-axis. Hence, they are in the regime of mixed properties. The fundamental role of the parameter γ becomes even more evident in such a representation: it determines the position of the beginning of the measured range and hence "shifts" the curves lying on the same hyperbola, e.g. 1 and 5 or 2 and 4, to the right, i.e. towards substrate, or to the left, i.e. towards polymer.

The film thickness t_f is given by [91]

$$t_f = c\frac{\varepsilon}{\alpha}\frac{\gamma - (\beta - \alpha)}{(\beta + \alpha) - \gamma} = c\frac{\varepsilon}{\alpha}\frac{\varepsilon/\alpha + \delta_m}{\varepsilon/\alpha - \delta_m}, \tag{1.120}$$

where c is a proportionality constant.

Later on the authors have found that Eq. (1.120) is not adequate for very large ranges of polymer thicknesses. In particular, the thickness is not estimated properly when $\gamma \approx \beta + \alpha$, i.e. for very thick polymer films. Hence, the expression of the polymer film thickness has been corrected in the form [92]

$$t_f = c\sqrt{\frac{\varepsilon}{\alpha}\frac{\gamma - (\beta - \alpha)}{(\beta + \alpha) - \gamma}}. \tag{1.121}$$

References

1. Binnig G, Quate CF, Gerber C (1986) Atomic force microscope. Phys Rev Lett 56:930–933
2. Binnig G, Rohrer H, Gerber C, Weibel E (1982) Surface studies by scanning tunneling microscopy. Phys Rev Lett 49:57–61
3. Lewis A, Isaacson M, Harootunian A, Muray A (1984) Development of a 500 Å spatial resolution light microscope. Ultramicroscopy 13:227–232
4. Pohl DW, Denk W, Lanz M (1984) Optical stethoscopy – Image recording with resolution lambda/20. Appl Phys Lett 44:651–653
5. Tortonese M (1997) Cantilevers and tips for atomic force microscopy. IEEE Eng Med Bio 16:28–33
6. Fleming AJ (2013) A review of nanometer resolution position sensors: operation and performance. Sensors Actuators A: Phys 190:106–126
7. Meyer G, Amer NM (1988) Novel optical approach to atomic force microscopy. Appl Phys Lett 53:1045–1047
8. Bhushan B (ed) (2004) Springer handbook of nanotechnology. Springer, Berlin
9. Schönherr H, Vancso GJ (2010) Scanning force microscopy of polymers. Springer, Heidelberg
10. Hamada E, Kaneko R (1992) Micro-tribological evaluations of a polymer surface by atomic force microscopes. Ultramicroscopy 42:184–190
11. Cappella B, Dietler G (1999) Force-distance curves by atomic force microscopy. Surf Sci Rep 34:1–104
12. Butt H-J, Cappella B, Kappl M (2005) Force measurements with the atomic force microscope: technique, interpretation and applications. Surf Sci Rep 59:1–152
13. Ferry JD (1961) Viscoelastic properties of polymers. Wiley, New York
14. Aimé JP, Elkaakour Z, Odin C, Bouhacina T, Michel D, Curély J, Dautant A (1994) Comments on the use of the force mode in atomic force microscopy for polymer films. J Appl Phys 76:754–762
15. Hertz H (1881) Über die Berührung fester elastischer Körper. J Reine Angew Math 92:156–171
16. Johnson KL, Kendall K, Roberts AD (1971) Surface energy and the contact of elastic solids. Proc R Soc Lond A 324:301–313
17. Derjaguin BV, Müller VM, Toporov YP (1975) Effect of contact deformations on the adhesion of particles. J Colloid Interf Sci 53:314–326
18. Müller VM, Yushchenko VS, Derjaguin BV (1980) On the influence of molecular forces on the deformation of an elastic sphere and its sticking to a rigid plane. J Colloid Interf Sci 77:91–101
19. Müller VM, Derjaguin BV, Toporov YP (1983) On two methods of calculation of the force of sticking of an elastic sphere to a rigid plane. Colloids Surf 7:251–259
20. Sneddon IN (1965) The relation between load and penetration in the axisymmetric Boussinesq problem for a punch of arbitrary profile. Int J Engng Sci 3:47–57
21. Attard P, Parker JL (1992) Deformation and adhesion of elastic bodies in contact. Phys Rev A 46:7959–7971
22. Pashley MD, Pethica JB, Tabor D (1984) Adhesion and micromechanical properties of metal surfaces. Wear 100:7–31

23. Maugis D (1992) Adhesion of spheres: the JKR-DMT transition using a Dugdale model. J Colloid Interf Sci 150:243–269
24. Maugis D (1999) Contact, adhesion and rupture of elastic solids. Springer, Berlin
25. Johnson KL (2000) Contact mechanics and adhesion of viscoelastic spheres. ACS Symp Ser 741:24–41
26. Oliver WC, Pharr GM (1992) An improved technique for determining hardness and elastic modulus using load and displacement sensing indentation experiments. J Mater Res 7:1564–1583
27. Briscoe BJ, Fiori L, Pelillo E (1998) Nano-indentation of polymeric surfaces. J Phys D Appl Phys 31:2395–2405
28. King RB (1987) Elastic analysis of some punch problems for a layered medium. Int J Solids Struct 23:1657–1664
29. Pharr GM, Oliver WC, Brotzen FR (1992) On the generality of the relationship among contact stiffness, contact area, and elastic-modulus during indentation. J Mater Res 7:613–617
30. Van Landingham MR, Chang NK, Drzal PL, White CC, Chang SH (2005) Viscoelastic characterization of polymers using instrumented indentation. I. Quasi-static testing. J Polym Sci B: Polym Phys 43:1794–1811
31. Balasundaram K, Cao Y, Raabe D (2008) Identifying the limitation of Oliver and Pharr method in characterizing the viscoelastic-plastic materials with respect to indenter geometry. Mater Res Soc Symp Proc 1137:49–54
32. Tranchida D, Piccarolo S, Loos J, Alexeev A (2007) Mechanical characterization of polymers on a nanometer scale through nanoindentation. A study of pile-up and viscoelasticity. Macromolecules 40:1259–1267
33. Israelachvili J (1992) Intermolecular and surface force. Academic Press, London
34. Gao C (1997) Theory of menisci and its applications. Appl Phys Lett 71:1801–1803
35. Eastman T, Zhu D-M (1996) Adhesion forces between surface-modified AFM tips and a mica surface. Langmuir 12:2859–2862
36. Weisenhorn AL, Hansma PK, Albrecht TR, Quate CF (1989) Forces in atomic force microscopy in air and water. Appl Phys Lett 54:2651–2653
37. Butt H-J (1991) Electrostatic interaction in atomic force microscopy. Biophys J 60:777–785
38. Gillies G, Prestidge GA (2005) Colloid Probe AFM investigation of the influence of cross-linking on the interaction behavior and nano-rheology of colloidal droplets. Langmuir 21:12342–12347
39. Dimitriadis EK, Horkay F, Maresca J, Kachar B, Chadwick RS (2002) Determination of elastic moduli of thin layers of soft material using the atomic force microscope. Biophys J 82:2798–2810
40. Derjaguin BV, Landau LD (1941) Theory of the stability of strongly charged lyophobic sols and of the adhesion of strongly charged particles in solution of electrolytes. Acta Phys URSS 14:633–662
41. Verwey EJW, Overbeek JTG (1948) Theory of stability of lyophobic colloids. Elsevier, Amsterdam
42. O'Shea SJ, Welland ME, Pethica JB (1994) Atomic-force microscopy of local compliance at solid–liquid interfaces. Chem Phys Lett 223:336–340
43. Butt H-J (1991) Measuring electrostatic, van der Waals, and hydration forces in electrolyte solutions with an atomic force microscope. Biophys J 60:1438–1444
44. Rabinovich YI, Yoon RH (1994) Use of atomic-force microscope for the measurements of hydrophobic forces between silanated silica plate and glass sphere. Langmuir 10:1903–1909
45. Milner ST, Witten TA, Cates ME (1988) A parabolic density profile for grafted polymers. Europhys Lett 5:413–418
46. Milner ST, Witten TA, Cates ME (1988) Theory of the grafted polymer brush. Macromolecules 21:2610–2619
47. Dolan AK, Edwards SF (1974) Theory of stabilization of colloids by adsorbed polymer. Proc R Soc Lond A 337:509–516

48. Fleer GJ, Scheutjens JHMH, Vincent B (1984) The stability of dispersions of hard spherical particles in the presence of nonadsorbing polymer. ACS Symp Ser 240:245–263
49. Biggs S (1995) Steric and bridging forces between surfaces bearing adsorbed polymer – an atomic-force microscopy study. Langmuir 11:156–162
50. Plazek DJ, Ngai KL (1996) The glass temperature. In: Mark JE (ed) Physical properties of polymers handbook. AIP Press, New York
51. Doi M (1993) Viscoelastic and rheological properties. In: Cahn RW, Haasen P, Kramer EJ (eds) Structure and properties of polymers. Wiley, Weinheim
52. Fried JR (1996) Sub-T_g transitions. In: Mark JE (ed) Physical properties of polymers handbook. AIP Press, New York
53. Peyser P (1989) Glass transition temperature of polymers. In: Brandrup J, Immergut EH (eds) Polymer handbook. AIP Press, New York
54. Fox T, Flory P (1954) The glass temperature and related properties of polystyrene – influence of molecular weight. J Polym Sci 14:315–319
55. Rubinstein M, Colby RH (2003) Polymer physics. Oxford University Press, Oxford
56. Williams ML, Landel RF, Ferry JD (1955) Mechanical properties of substances of high molecular weight. The temperature dependence of relaxation mechanisms in amorphous polymers and other glass-forming liquids. J Am Chem Soc 77:3701–3707
57. Chyasnavichyus M, Young SL, Tsukruk VV (2014) Probing of polymer surfaces in the viscoelastic regime. Langmuir 30:10566–10582
58. Cappella B, Kaliappan SK, Sturm H (2005) Using AFM force distance curves to study the glass-to-rubber transition of amorphous polymers and their elastic–plastic properties as a function of temperature. Macromolecules 38:1874–1881
59. Cohen SR, Kalfon-Cohen E (2013) Dynamic indentation by instrumented nanoindentation and force microscopy: a comparative review. Beilstein J Nanotechnol 4:815–833
60. Lu H, Wang B, Ma J, Huang G, Viswanathan H (2003) Measurement of creep compliance of solid polymers by nanoindentation. Mech Time-Depend Mater 7:189–207
61. Jäger A, Lackner R (2009) Finer-scale extraction of viscoelastic properties from nanoindentation characterised by viscoelastic-plastic response. Strain 45:45–54
62. Braunsmann C, Proksch R, Revenko I, Schäffer TE (2014) Creep compliance mapping by atomic force microscopy. Polymer 55:219–225
63. Asif SAS, Wahl KJ, Colton RJ (1999) Nanoindentation and contact stiffness measurement using force modulation with a capacitive load–displacement transducer. Rev Sci Instrum 70:2408–2413
64. Asif SAS, Wahl KJ, Colton RJ, Warren OL (2001) Quantitative imaging of nanoscale mechanical properties using hybrid nanoindentation and force modulation. J Appl Phys 90:1192–1200
65. Burnham NA, Gremaud G, Kulik AJ, Gallo PJ, Oulevey F (1996) Materials' properties measurements: choosing the optimal scanning probe microscope configuration. J Vac Sci Technol B 14:1308–1312
66. Nalam PC, Gosvami NN, Caporizzo MA, Russell J, Composto RJ, Carpick RW (2015) Nano-rheology of hydrogels using direct drive force modulation atomic force microscopy. Soft Mater 11:8165–8178
67. Odegard GM, Gates TS, Herring HM (2005) Characterization of viscoelastic properties of polymeric materials through nanoindentation. Exp Mech 45:130–136
68. White CC, VanLandingham MR, Drzal PL, Chang NK, Chang SH (2005) Viscoelastic characterization of polymers using instrumented indentation. II Dynamic testing. J Pol Sci B: Pol Phys 43:1812–1824
69. Chakravartula A, Komvopoulos K (2006) Viscoelastic properties of polymer surfaces investigated by nanoscale dynamic mechanical analysis. Appl Phys Lett 88:131901
70. Bouaita N, Bull SJ, Fernandez Palacio J, White JR (2006) Dynamic nanoindentation of some polyolefins. Pol Eng Sci 46:1160–1172

71. Zhou J, Komvopoulos K (2007) Interfacial viscoelasticity of thin polymer films studied by nanoscale dynamic mechanical analysis. Appl Phys Lett 90:021910
72. Zhang Y-F, Bai S-L, Yang D-Y, Zhang Z, Kao-Walter S (2008) Study on viscoelastic properties of the epoxy surface by means of nanodynamic mechanical analysis. J Pol Sci B: Pol Phys 46:281–288
73. Lu YC, Shinozaki DM (2010) Temperature dependent viscoelastic properties of polymers investigated by small-scale dynamic mechanical analysis. Exp Mech 50:71–77
74. Takahashi R, Okajima T (2015) Mapping power-law rheology of living cells using multi-frequency force modulation atomic force microscopy. Appl Phys Lett 107:173702
75. Hecht FM, Rheinlaender J, Schierbaum N, Goldmann WH, Fabry B, Schäffer TE (2015) Imaging viscoelastic properties of live cells by AFM: power-law rheology on the nanoscale. Soft Mater 11:4584–4591
76. Nakajima K, Ito M, Wang D, Liu H, Nguyen HK, Liang X, Kumagai A, Fujinami S (2014) Nano-palpation AFM and its quantitative mechanical property mapping. Microscopy 63:193–207
77. Igarashi T, Fujinami S, Nishi T, Asao N, Nakajima K (2013) Nanorheological mapping of rubbers by atomic force microscopy. Macromolecules 46:1916–1922
78. Tranchida D, Kiflie Z, Piccarolo S (2007) Viscoelastic recovery behavior following atomic force microscope nanoindentation of semicrystalline poly(ethylene). Macromolecules 40:7366–7371
79. Cappella B, Sturm H (2002) Comparison between dynamic plowing lithography and nanoindentation methods. J Appl Phys 91:506–512
80. Du B, Tsui OKC, Zhang Q, He T (2001) Study of elastic modulus and yield strength of polymer thin films using atomic force microscopy. Langmuir 17:3286–3291
81. Johnson KL (1970) The correlation of indentation experiments. J Mech Phys Solids 18:115–126
82. Mencik J, Munz D, Quandt E, Weppelmann ER (1997) Determination of elastic modulus of thin layers using nanoindentation. J Mater Res 12:2475–2484
83. Perriot A, Barthel E (2004) Elastic contact to a coated half-space: effective elastic modulus and real penetration. J Mater Res 19:600–608
84. Jung Y-G, Lawn BR, Martyniuk M, Huang H, Hu XZ (2004) Evaluation of elastic modulus and hardness of thin films by nanoindentation. J Mater Res 19:3076–3080
85. Clifford CA, Seah MP (2006) Modelling of nanomechanical nanoindentation measurements using an AFM or nanoindenter for compliant layers on stiffer substrates. Nanotechnology 17:5283–5292
86. Tsukruk VV, Sidorenko A, Gorbunov VV, Chizhik SA (2001) Surface nanomechanical properties of polymer nanocomposite layers. Langmuir 17:6715–6719
87. Kovalev A, Shula H, Lemieux M, Myshkin N, Tsukruk VV (2004) Nanomechanical probing of layered nanoscale polymer films with atomic force microscopy. J Mater Res 19:716–728
88. Doerner MF, Nix WD (1986) A method for interpreting the data from depth-sensing indentation instruments. J Mater Res 1:601–609
89. Gao H, Chiu CH, Lee J (1992) Elastic contact versus indentation modeling of multilayered materials. Int J Solids Struct 29:2471–2492
90. Cappella B, Silbernagl D (2008) Nanomechanical properties of polymer thin films measured by force–distance curves. Thin Solid Films 516:1952–1960
91. Cappella B, Silbernagl D (2007) Nanomechanical properties of mechanical double-layers: a novel semiempirical analysis. Langmuir 23:10779–10787
92. Silbernagl D, Cappella B (2010) Mechanical properties of thin polymer films on stiff substrates. Scanning 32:282–293

Force–Distance Curves in Practice

2

Abstract

In this chapter, basic experimental aspects of the use of an atomic force microscope for the acquisition of force–distance curves and the study of mechanical properties of samples are discussed.

In the first two sections, calibration issues (sensitivity, spring constant of the cantilever and radius of the cantilever tip) are treated; also, the colloidal probe technique is briefly presented, and advantages and drawbacks are discussed.

In Sect. 2.3 fundamental aspects of data analysis for force–distance curves are described. Moreover, the most common artefacts affecting the acquisition and the analysis of force–distance curves and in particular of deformation–force curves are listed.

Section 2.4 summarizes in table form the sequence of work steps of an experiment aimed to the measurement of mechanical properties of the sample through force–distance curves.

2.1 Optical Lever Technique and Sensitivity

The cantilever deflection is usually measured using the optical lever technique [1]. In this technique, a laser beam is focused onto the end of the cantilever, and the position of the reflected spot is measured with a position-sensitive detector, commonly a four-quadrant photodiode.

In order to relate the displacement of the reflected spot to the deflection, it is necessary at first to calculate the deflection of a rectangular cantilever of length L_c, whose shape is described by the function $Z(X)$, being X the position along the cantilever fixed at $X = 0$. The torque due to the force F at a given position X is $F(L_c - X)$. If E_c is the elastic modulus of the cantilever and I_c its moment of inertia, the following differential equation can be written [2]:

B. Cappella, *Mechanical Properties of Polymers Measured through AFM Force-Distance Curves*, Springer Laboratory, DOI 10.1007/978-3-319-29459-9_2

$$F(L_c - X) = E_c I_c \frac{\partial^2 Z}{\partial X^2}. \tag{2.1}$$

If the cross section of the cantilever is constant, I_c does not depend on X. With the boundary conditions $Z(X=0)=0$ and $dZ/dX(X=0)=0$, the solution of Eq. (2.1) is

$$Z = \frac{F}{2E_c I_c}\left(L_c X^2 - \frac{X^3}{3}\right). \tag{2.2}$$

The deflection is calculated by putting $X = L_c$:

$$\delta = \frac{FL_c^3}{3E_c I_c}. \tag{2.3}$$

The reflected laser beam moves through an angle α equal to twice the change of the end slope of the cantilever.

$$\alpha = 2\frac{\partial Z}{\partial X}\bigg|_{L_c} = \frac{FL_c^2}{E_c I_c} = \frac{3\delta}{L_c}. \tag{2.4}$$

If the four-quadrant photodiode is at a distance d from the cantilever, the displacement of the reflected spot Δ_S is

$$\Delta_S = d \tan \alpha \cong d\alpha = \frac{dFL_c^2}{E_c I_c} = \frac{3d}{L_c}\delta. \tag{2.5}$$

Given a certain experimental setup, the displacement of the reflected spot Δ_S is hence proportional to the cantilever deflection δ.

The displacement of the reflected spot Δ_S is measured by the four-quadrant photodiode as the voltage $\Delta V = V_{top} - V_{bottom}$, where V_{top} and V_{bottom} are the sum of the voltages detected in the top and bottom quadrants, respectively. The cantilever deflection is given by $\delta = \Delta V/\Omega$, where the parameter Ω is the sensitivity of the system, i.e. the conversion factor between the deflection in volts and the deflection in metres.

As already seen in Sect. 1.2, the relation between the cantilever deflection and the piezo displacement is given by $\delta = (k_{eff}/k_c)Z$, where $k_{eff} = k_c k_s/(k_c + k_s)$. If $k_s \gg k_c$, $k_{eff} \approx k_c$ and $\delta \approx Z$. This means that a deflection–displacement curve on a very stiff sample can be used to determine the sensitivity Ω, since Z is measured in metres.

Therefore, to calibrate the optical lever system, after adjusting the position of the photodiode, so that the deflection is zero, a deflection–displacement curve is acquired on a very stiff sample, usually a silicon or glass slide, and the contact line of the curve is fitted with a straight line. Putting the slope of the fitting line equal to one yields the sensitivity Ω.

The sensitivity must be measured at the beginning of every measurement. It should be kept in mind that the sensitivity may change during a very long measurement. Therefore, if it is possible to interrupt the measurement and to change the sample, the value of Ω should be corrected at regular time intervals during the measurement.

The assumption that δ and Z are the same for a very stiff sample may be a too rough approximation, depending on the aim of the measurement. When a resolution of about 1 nm in the measurement of the deformation is required, Ω must be determined with a higher precision, and the deformation of silicon or glass cannot be ignored. For example, the deformation of a silicon sample ($E \approx 150$ GPa) indented with a silicon nitride tip ($E = 310$ GPa) of radius $R = 25$ nm and with a cantilever of elastic constant $k_c = 40$ N/m is about 3 nm already at a deflection $\delta = 100$ nm.

In this case the deformation in volts can be fitted with the Hertz equation in the form:

$$D = \left(\frac{k_c}{E_{\text{tot}} \sqrt{R}} \frac{\Delta V}{\Omega} \right)^{2/3}. \tag{2.6}$$

Since the elastic modulus is known, the fit yields the value of Ω.

2.2 AFM Cantilevers and Tips

2.2.1 Determination of the Elastic Constant

Once the sensitivity Ω is determined, the deflection in metre is known. Yet, in all elastic continuum theories, knowledge of the force is required. In order to calculate the force from the deflection through Hooke's law (Eq. 1.1), the elastic constant of the cantilever, k_c, is needed.

The spring constant of a rectangular cantilever can be calculated using Eq. (2.3):

$$k_c = \frac{F}{\delta} = \frac{3E_c I_c}{L_c^3} = \frac{E_c w_c t_c^3}{4L_c^3}, \tag{2.7}$$

where $I_c = w_c t_c^3/12$ has been factorized.

The spring constant of V-shaped cantilevers can be approximated with the constant of a rectangular cantilever of width $2w_c$. A most accurate expression has been calculated by Neumeister and Ducker [3]:

$$k_c = \left[\Delta_1 + \Delta_2 + \Psi \left(\frac{w_c}{\sin \alpha} - d_c \right) \right]^{-1}, \tag{2.8}$$

with

$$\Delta_1 = \frac{3}{Et_c^3 \tan\alpha_c}\left[\left(\frac{w_c}{\sin\alpha_c} - 2d_c\right)^2 - d_c^2\left(2\log\frac{w_c}{d_c \sin\alpha_c} + 1\right)\right], \tag{2.9}$$

$$\Delta_2 = \frac{L_c^2}{Ew_c t_c^3 \cos^2\alpha_c}\left[\frac{2L_c}{\cos\alpha_c} + 3(w_c \cot\alpha_c - d_c \cos\alpha_c - \sin\alpha_c)\right], \tag{2.10}$$

$$\Psi = \frac{3L_c(1+\nu)}{Ew_c t_c^3 \cos\alpha_c}\left(\frac{w_c}{\sin\alpha_c} - d_c + \Theta\cos\alpha_c\right) \tag{2.11}$$

and

$$\Theta = \frac{L_c \tan\alpha_c + (w_c - d_c \sin\alpha_c)(1-\nu)\cos\alpha_c}{2 - (1-\nu)\cos^2\alpha_c}. \tag{2.12}$$

The geometrical quantities employed in the previous equations are defined in Fig. 2.1.

For the calculation of the elastic constant through Eqs. (2.7) or (2.8), a precise knowledge of several geometrical and mechanical quantities is necessary. This may be rather challenging in case of the elastic modulus E, of Poisson's ratio ν and of the thickness t_c. This is a limitation also for calculations based on finite element modelling of the cantilever [3].

Values of the elastic modulus of materials commonly used for the fabrication of AFM cantilevers, necessary also for the calculation of the reduced elastic modulus (see Eq. 1.12 in Sect. 1.4), are listed in Table 2.1.

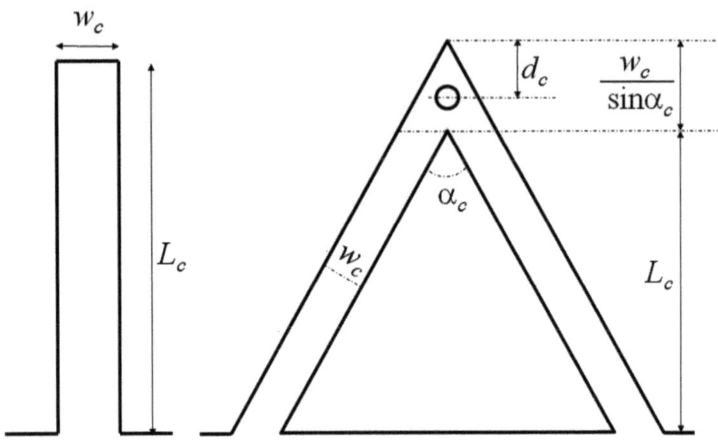

Fig. 2.1 Geometry of a rectangular and a V-shaped cantilever. For the rectangular cantilever, L_c and w_c are the length and the width. For the V-shaped cantilever, $L_c + w_c/\sin\alpha_c$ is the total length, w_c is the width of the arms, α_c is the angle between the arms and d_c is the distance between the centre of the tip and the end of the cantilever

Table 2.1 Elastic modulus E and Poisson's ratio ν of materials commonly used for the fabrication of AFM cantilevers

Material	Young's modulus E (GPa)	Poisson's ratio ν
Diamond	900–1050	0.07
Si	130–188	0.28
Si_3N_4	160–310	0.27
W	350	0.28
Ir	530	0.26

Even if all geometrical and mechanical quantities are known, several factors lead to significant deviations of the actual value of the spring constant from the theoretical values obtained through Eqs. (2.7) or (2.8). One major factor is that the thickness of the cantilever is usually not homogeneous; even small deviations of t_c lead to significant deviations of k_c. Another factor is the presence of a metal layer on top of the cantilever, changing its mechanical properties.

Because of the numerous error sources, the elastic constant of a cantilever should be determined experimentally. To this aim, several methods have been developed in the past decades.

One of the most reliable methods is to use a reference cantilever [4] or a spring [5] of known spring constant k_{ref}. The reference cantilever is mounted on a piezo translator and pushed against the cantilever, whose spring constant has to be determined. The spring constant is given by

$$k_c = k_{\text{ref}} \frac{Z - \delta}{\delta} = k_{\text{ref}} \frac{1 - \delta/Z}{\delta/Z}, \tag{2.13}$$

where Z is the piezo displacement and δ the cantilever displacement.

The elastic constant of a cantilever can be determined also by means of a nanoindenter [6]. This method is suited only for stiff cantilevers ($k_c > 1$ N/m) and is limited by the difficulty of placing the indenter tip in the right position onto the AFM cantilever.

A third method [7] consists in measuring the resonance frequency of a cantilever with and without a known extra mass M attached to it.

The resonance frequencies of a cantilever of mass m, $\omega_0 = \sqrt{k_c/m}$, change to $\omega_1 = \sqrt{k_c/(m + M)}$ when the mass M is attached to it. Hence

$$k_c = \frac{M}{1/\omega_1^2 - 1/\omega_0^2}. \tag{2.14}$$

The precision of this method depends crucially on the exact determination of the (small) mass M. Furthermore, since the method must be nondestructive, the mass M must be attached through capillary force and not glued onto the cantilever.

The easiest method for the measurement of k_c is that proposed by Hutter and Bechhoefer [8]. In this method the power spectral density of the deflection of a cantilever due to thermal noise is measured. If the cantilever is considered as a harmonic oscillator, then

$$\left\langle \frac{1}{2} m \omega_0^2 \delta^2 \right\rangle = \frac{1}{2} k_B T \tag{2.15}$$

and, since $\omega_0^2 = k_c/m$,

$$k_c = \frac{k_B T}{\langle \delta^2 \rangle}. \tag{2.16}$$

This method is the most commonly used, because it does not require any equipment (reference cantilever or attached extra mass) and is implemented in most commercial microscopes. In practice, after measuring the sensitivity Ω (see Sect. 2.1) in order to know the deflection in metres, a noise spectrum of the deflection amplitude is acquired. The peak at the resonance frequency is fitted with a Lorentz curve, and the mean square deflection is obtained via integration.

Errors due to multiple vibration modes can be taken into account with a correction factor [9]. Anyway, the major source of error in this method is the precise determination of the deflection amplitude in metres. Since the deflection due to thermal noise is usually below 1 nm, the sensitivity Ω must be determined with very high precision. Uncertainties in the value of Ω affect strongly the calibration of the elastic constant of compliant cantilevers ($k_c < 1$ N/m). Furthermore, when the cantilever is very stiff, the deflection due to thermal noise is well under 1 nm and very difficult to detect.

A review of different calibration methods with a list of advantages and limitations can be found in [10].

2.2.2 Determination of the Tip Radius

Together with the spring constant k_c, the shape of the tip is a very important factor for the quantitative measurement of mechanical properties.

First of all, for the application of an elastic continuum theory, the knowledge of the shape of the tip apex is necessary. Furthermore, quantitative results can be obtained only if geometrical parameters (e.g. the radius in case of a spherical tip) are known with nanometre resolution.

AFM tips with a spherical apex have usually a radius R of some tens of nanometres down to 5 nm. Hence, a precise determination of R is rather challenging.

The characterization of the tip shape is often achieved by means of a scanning electron microscope (SEM). Yet, it should be kept in mind that a SEM does not provide a direct information about the Z direction and that the three-dimensional shape must be reconstructed with the help of algorithms. Furthermore, the tip, which is usually insulating, must be coated with a layer of conducting material. The resolution is so limited to typically 10 nm.

A more effective method, which can be performed with the AFM itself, is to image a very sharp structure with the tip of unknown shape. The obtained image is actually the convolution of the profiles of the tip and of the sample structure. Hence, the tip shape can be reconstructed from such an image [11, 12].

Such measurements can be performed with commercial available test gratings with sharp tips. Errors in the determined tip radius are due to the finite size of the tips on the test grating. Unfortunately, since the test gratings are usually made out of silicon, the tip is likely to be damaged during the scan on the grid and to become dull.

Figure 2.2 shows a 3D image of the tip on a test grating (TGT1, NT-MDT, Moscow, Russia) obtained with a commercial AFM tip. The area of the image is 800×800 nm^2. Since the tips on the grating are about 500 nm high, only the last 450–500 nm of the AFM tip can be detected. The tip is pyramidal, but the apex has a parabolic section.

Figure 2.3 shows the vertical cross section over the apex of the replica of the tip. The apex is magnified in the inset, which shows also the fit with the function of a circle, $y = y_0 + \sqrt{R_{\text{fit}}^2 - (x - x_0)^2}$. The fit should be repeated for several sections of the tip. In the present case, the parameter R_{fit} is 35 nm.

The parameter R_{fit} is the maximum radius of the tip apex. As a matter of fact, the radius cannot be larger than R_{fit}, but, due to the finite size of the tips of the grating, it is certainly smaller. Assuming that the grating tips have a mean curvature radius R_{gr}

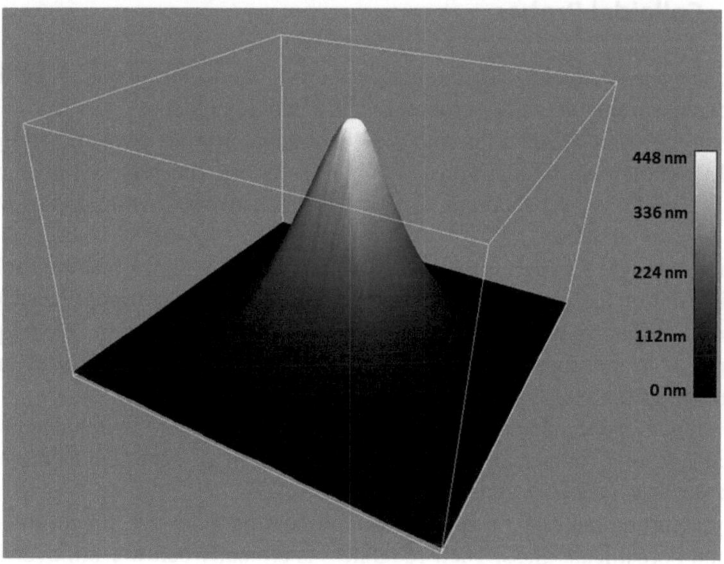

Fig. 2.2 3D image obtained on a test grating with a commercial tip. The area of the image is 800×800 nm^2

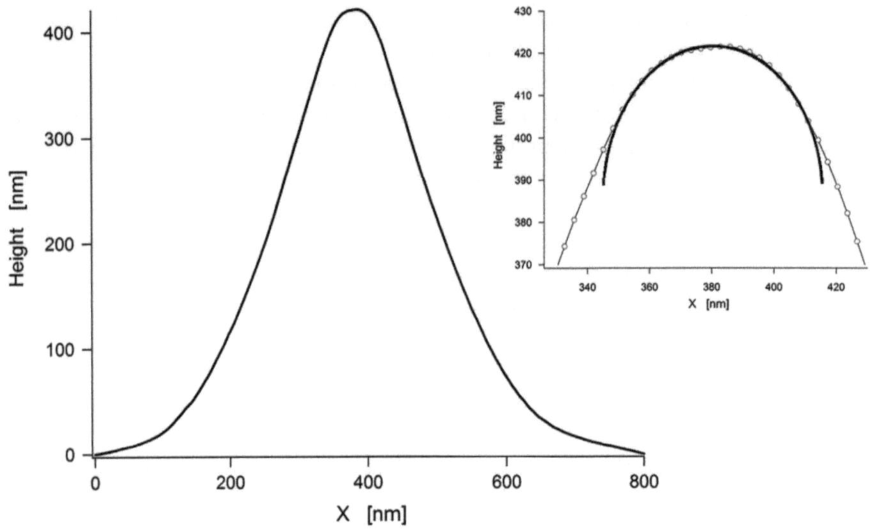

Fig. 2.3 Vertical cross section over the apex of the replica of the tip. The inset shows the very apex and the fit with a circle function

(in this case 5 nm), the apex radius R is given by $R = (R_{\mathrm{fit}} - R_{\mathrm{gr}}) \pm R_{\mathrm{gr}}$. In this case is hence $R = 30 \pm 5$ nm.

2.2.3 Colloidal Probes

A common way to avoid uncertainties and errors due to the shape of the tip is to perform measurements with a probe of well-defined geometry.

A first approach to use a tip with a defined geometry has been shown by Hüttl et al. [13]. In their work, common silicon AFM tips have been etched in an oven in the presence of oxygen, obtaining tips with a spherical apex with a radius of some hundreds of nanometres.

A more feasible solution is to use a colloidal particle with well-defined spherical shape instead of the micro-fabricated tip. The use of such colloidal particles was introduced by Ducker et al. [14, 15] and Butt [16]. Since then, this technique has been applied in the study of surface forces, but also in measurements dealing with mechanical properties.

The choice of the material to be used as colloidal particle is influenced by the fact that not every material is available in the form of spherical particles with smooth surfaces (see [17] and references therein).

Very common materials employed as colloidal particles are silica (amorphous SiO_2) and borosilicate glass. Their roughness is typically below 1 nm over 1 mm^2, and they are commercially available at different sizes. Moreover, the surface of such particles can be easily modified and functionalized. Other materials used are zirconia, alumina (Al_2O_3), titanium oxide (TiO_2), magnesium oxide (MgO), zinc and gold. Also polymeric microspheres made of polystyrene, poly(methyl methacrylate) or polyethylene are commercially available.

Colloidal probes must be attached to the end of a tipless cantilever. Usually this is achieved by gluing the particles with epoxy resins or UV-curable glues. To this aim, a small amount of glue is put onto the end of the cantilever, and successively colloidal probe and cantilever are brought in contact. This is done by means of a micromanipulator controlled through an optical microscope. Glue and particles can be moved with tiny wires or can be picked up by the cantilever itself. Depending on the material, colloidal particles can be sintered to the cantilever. Yet, this is not feasible when the particles are made of silica, due to the high melting temperature (1723 °C).

Once the colloidal probe has been attached, its radius must be measured. This can be achieved with the help of commercial gratings, as exposed in the previous section. Due to the considerably larger dimensions of a colloidal probe compared with a conventional AFM tip, the characterization of the geometry can be performed also by means of SEM images.

Independently of the materials, colloidal particle have typically a radius between 1 and 50 μm. Such large radii are the origin of the two main drawbacks of the colloidal probe technique, balancing the advantage of performing experiments with defined probe geometry.

The first drawback concerns force–volume measurements, where the study of the spatial variation of mechanical properties is aimed. As discussed later in Sect. 2.3.1, the resolution of a force–volume measurement, like the resolution in contact mode or in tapping mode, depends also on the tip radius. A very large tip radius, as in case of a colloidal tip, leads to a poor resolution.

The second drawback is important for measurements of mechanical properties. In this case the sample has to be deformed. In Hertz theory the dependence of the deformation on the force is given by $D = \sqrt[3]{\frac{k_c^2}{R}}\left(\frac{\delta}{E_{\text{tot}}}\right)^{2/3}$. Hence, if the same deformation is aimed in a measurement with a colloidal probe of radius $R = 10$ μm and in a measurement with a conventional AFM tip with a spherical apex of radius $R = 10$ nm, the elastic constant of the cantilever bearing the colloidal particle should be larger than the other one by a factor of ca. 30. Since commercially available cantilevers have an upper limit of the elastic constant of about 100 N/m, in some cases the deformations obtained with colloidal probes are too small for a meaningful fit of the deformation–force curve.

2.3 Practical Issues of Force–Distance Curves Acquisition

2.3.1 Data Analysis and Force–Volume Measurements

Data Analysis

Any information about the mechanical behaviour of the sample is derived from the contact lines of force–distance curves. Except when dealing with the sample

stiffness k_s or with the effective stiffness k_{eff}, measurement results are often presented in the form of deformation–force curves rather than force–distance curves by calculating the deformation through Eq. (1.2), $D = Z–\delta$. Subsequently, the deformation–force curve is fitted with one of the continuum elastic theories presented in Sect. 1.4. Since in Hertz and DMT theory the force is proportional to the deformation to the power of 3/2, also $D^{3/2}$ curves, i.e. the functions $D^{3/2}(F)$ or $D^{3/2}(\delta)$, are often presented as measurement results.
Essential requirements of the analysis of a force–distance curve are:

1. Identification of the discontinuities (jump-to-contact and jump-off-contact) or, when no discontinuities are present, of the points at which contact is established or lost. This step is necessary to separate the zero line and the contact region.
2. Fit of the zero line, yielding the zero of the Y axis (force or deflection).
3. Identification of the zero of the X-axis (piezo elongation).

When only mean quantities of a homogeneous sample, e.g. Young's modulus, are of interest, not only the measured quantities but also the whole force–distance cycle should be averaged over tens of single curves. This is due to the fact that a single force–distance curve is strongly influenced by the topography of the sample at the point where the curve has been acquired. The topography of the sample is never ideally flat and, even on very compliant samples, small asperities change the contact radius and consequently at least the initial part of the contact line (see next section).

When single force–distance curves are to be averaged, they must be referred to the same origin. Hence, automated procedures encompassing at least the three just mentioned steps of the analysis are necessary.

The easiest way to identify the jump-to-contact and the jump-off-contact is to seek the minima of the approach and retraction curve. Yet, in some cases, the zero line has a nonzero slope, and the discontinuities are not the minima of the curves. Also, attractive and adhesive forces can be so small that discontinuities are smaller than the average noise along the zero line.

The jump-to-contact and the jump-off-contact can be found by seeking the minimum of the first derivative of the force with respect to the piezo displacement as shown schematically in the left part of Fig. 2.4. The left top panel shows a force–displacement curve and the bottom panel its first derivative dF/dZ. It is $\mathrm{d}F/\mathrm{d}Z = k_{\text{eff}} = k_c k_s/(k_c + k_s)$ along the contact line and $\mathrm{d}F/\mathrm{d}Z = 0$ along the zero line. The two regions with constant values are separated by two sharp minima at the position of the discontinuities.

The existence of a minimum in the first derivative at the point where contact is established or lost does not depend on the assumption that the sample behaves as an ideal spring with elastic constant k_s, and such an analysis can be performed even when the deformation is described by one of the elastic continuum theories. Yet, this procedure is feasible only when the jump-to-contact and the jump-off-contact are pronounced discontinuities. This is usually the case of experiments performed

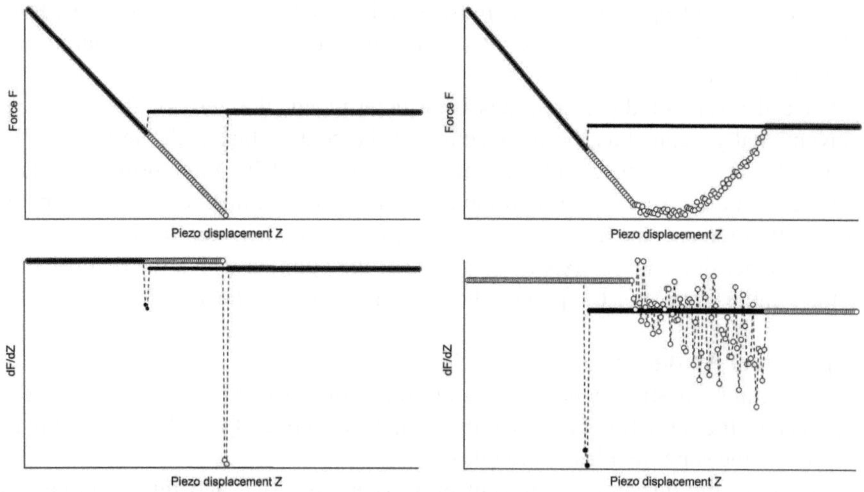

Fig. 2.4 Identification of the discontinuities through the first derivative. In the *top panels*, two schematic force–distance curves are shown. The approach curves are identical; the retraction curve in the *right panel* presents a slide-off-contact instead of a jump-off-contact. The first derivatives are shown in the *bottom panels*

in air, due to the capillary force. Nevertheless, even in air, if a neck is built when retracting the tip, the withdrawal curve may present a so-called slide-off-contact instead of a jump-off-contact, as shown in the right part of Fig. 2.4, where the approach curve is the same as before.

Since the slide-off-contact is not a discontinuity, the zero line and the contact line in the withdrawal curve are not separated by a sharp minimum, and the first derivative of the force gives no information about the point at which contact is lost. The situation is even more complicated in the presence of a repulsive force prior to contact, e.g. double-layer repulsion or steric repulsion (see Sect. 1.5). In this case the jump-to-contact and eventually also the jump-off-contact are neither a discontinuity nor a minimum. Furthermore, if the sample has a low elastic modulus, the repulsive force engenders deformations also before intimate contact is established. In this case there is no way to identify the point where the sample starts to be deformed, and even the analysis of single curves is rather challenging and requires ad hoc approaches [18–21].

Once the points where contact is established and lost have been identified, it is possible to separate the zero lines from the contact regions both in the approach and in the withdrawal curves.

In air, where almost no repulsive or attractive force prior to the jump-to-contact is present, the zero line is a straight line, but not always a horizontal line with zero slope. A nonzero slope may be engendered also in air by friction, especially with very compliant cantilevers. In general, a line $\delta = mZ + q$ must be subtracted to the deflection (or to the force) up to the jump-to-contact (jump-off-contact). The value

$\delta_{\text{jtc}} = mZ_{\text{jtc}} + q$ (or $\delta_{\text{joc}} = mZ_{\text{joc}} + q$) must be subtracted from the rest of the curve, where Z_{jtc} (Z_{joc}) is the piezo extension at which the jump-to-contact (jump-off-contact) occurs.

Once the zero of the Y axis has been determined, the zero of the X-axis is conventionally defined as the intersection of the contact line with the axis $Y = 0$. Again, this definition poses some problems when a repulsive force prior to contact is present. In this case only the extrapolation of the contact line, and not the contact line itself, intersects the axis $Y = 0$. Yet, the intersection between the extrapolation of the contact line and the axis $Y = 0$ is neither the point, at which the deformation of the sample starts, nor the point, where the tip 'touches' the sample surface.

Approach or Withdrawal Contact Line?

An important question when analysing force–distance curves is whether the approach or the retraction curve should be fitted with one of the elastic continuum theories to measure the elastic modulus.

Of course the question is relevant only when approach and retraction curves differ, i.e. when the loading–unloading cycle shows a hysteresis. The hysteresis may be due to adhesion, engendering a large difference between jump-to-contact and jump-off-contact (see Sect. 1.4.3), or to plastic deformations, causing a hysteresis loop in the contact region.

When the contact lines overlap and the only hysteresis is due to the difference between jump-to-contact and jump-off-contact, often the retraction contact line is fitted with DMT or JKR theory to get the work of adhesion W by putting the jump-off-contact equal to $2\pi RW$ (DMT) or $3\pi RW/2$ (JKR). Yet, if the microscope is operated in air, the jump-off-contact includes also a large contribution due to the capillary force. The capillary force is an interaction between the tip and the water film on top of the sample and not between the tip and the sample. In other words, the capillary force contributes to the deformation of the water film, but not to the deformation of the underlying sample. When the jump-off-contact is attributed only to tip-sample interactions, the adhesion is overestimated, and the modulus calculated with DMT or JKR theory is lower than the real modulus [22, 23].

When hysteresis is present also in the contact region (i.e. the contact lines do not overlap), plastic deformations as defined in Sect. 1.9 occur; the contact area is much larger than the one predicted by DMT or JKR theory, and putting the jump-off-contact equal to $2\pi RW$ or $3\pi RW/2$ leads to a severe error. Furthermore, plastic deformations affect also the retraction contact line, and its fit with DMT or JKR equation yields a false value of E_{tot}. The effect of plastic deformations is particularly relevant for very compliant polymers, since in this case plastic deformations cannot be avoided.

Finally, especially for compliant polymers, it has been shown [24–26] that the adhesion force depends also on the loading rate and on the contact time between probe and sample and that the retraction curve cannot be described by the DMT or JKR theory.

In particular, the work of adhesion W is a function of both the loading rate v_F and the contact time t_c:

$$W = W_0(t_c)\left[1 + \alpha v_p^n\right], \tag{2.17}$$

where $v_p = \mathrm{d}a/\mathrm{d}t$ is the displacement velocity of the contact line or the velocity of peeling off two surfaces, W_0 is the work of adhesion at $v_p = 0$ and α and n are material-specific constants.

The term αv_p^n depends on the dissipation energy at the border of the contact region during the detachment and affects the work of adhesion only when tip and sample surfaces are actually detached. Since v_p increases with v_F, also W and F_{adh} increase with v_F. The term $W_0(t_c)$ takes into account that the bond formation between surfaces in contact increases with the contact time. Hence, both W_0 and W increase with increasing t_c.

It is important to note that the dependence of the adhesion force on the loading rate which is discussed here, although following the time–temperature superposition principle, is not related to the deformation of the sample. As a matter of fact, the contact time is usually varied by introducing a so-called dwell time between approach and retraction; hence, it does not influence the deformation rate, since the sample is deformed before or after the dwell time, but not during the dwell time.

Because of the dependence illustrated in Eq. (2.17), unless v_p is zero, the work of adhesion measured from the jump-off-contact is not W_0; yet, the work of adhesion affecting the deformation and used in DMT or JKR equations is W_0.

Fitting the approach curve avoids both errors due to the capillary force and to plastic deformations, since the capillary force is an adhesive force and plastic deformations do not occur at the beginning of the loading process. Yet, when the adhesion is estimated from the jump-to-contact, the value of W is underestimated. This is due mainly to the fact that three important factors are ignored: the effect of bond formation, the inherent hysteresis of loading–unloading curves due to the elastic force of the cantilever (see Sect. 1.2) and the inherent hysteresis in JKR theory (see Sect. 1.4.3).

Summarizing, both portions of the force–distance curve yield a false estimate of the work of adhesion. The choice of the approach or of the retraction curve for fitting is hence a compromise, which depends on the particular experimental setup and on the material under study.

Force–Volume Measurements

One of the most advantageous features of an AFM is the capability of scanning the sample surface with high lateral resolution (atomic resolution on stiff samples). Scanning of a surface can be performed also to detect the lateral variation of force–distance curves, for example, to investigate the local distribution of the elastic moduli of a nonhomogeneous sample such as a polymer blend.

An array of force–distance curves on a sample surface is called *force–volume*. Usually force–distance curves in a force–volume are acquired with the same maximum force and the same velocity.

An important parameter in a force–volume is the *XY* distance or spacing between two successive curves, since it is related to the lateral resolution of the force–volume. Yet, the term ‘resolution’ should be employed carefully.

When measuring the elastic modulus of the sample, it is necessary to indent it. Furthermore, for a meaningful fit, the number of points of the deformation–force curve should be at least greater than ten and the spacing of the points should not be considerably smaller than 1 Å. Hence, the sample deformation must be at least some nanometres and cannot be reduced indefinitely. In addition, if the indentation is smaller than 1 nm, the fine structure of the sample must be considered; as a consequence, elastic continuum theories cannot be applied.

Even with a maximum indentation as small as $D_{max} = 1$ nm and a very sharp tip (R between 5 and 10 nm), the contact diameter $2a$ would be between 4 and 6 nm. More realistic values ($D_{max} = 5$ nm and $R = 15$ nm) yield $2a = 17$ nm.

The contact diameter $2a$ is a measure of the dimension of the 'probe' used for the detection of the contact line, for the calculation of the deformation and the determination of all related quantities.

Let us consider an ideal polymer blend with an infinitely sharp interface between the two phases with different elastic moduli; furthermore, suppose that there is no 'hard-wall' effect, i.e. the modulus of the sample changes abruptly, and not gradually, at the interface. In other words, the modulus as a function of the distance from the interface is a step function.

If we acquire a force–volume across such an interface with spacing between the curves smaller than $2a$, at a distance from the interface smaller than $2a$, the tip will sense a sample, whose elastic modulus is intermediate between those of the two phases. Hence, instead of a step function, the dependence of the measured elastic modulus on the distance from the interface is described by a sigmoidal function. In other words, the measured function is a convolution between the real step function and a $2a$-wide rectangular function.

The important consequence is that decreasing the spacing of the force–distance curves does not improve the resolution of a force–volume, which is limited by the contact radius.

2.3.2 Typical Artefacts of AFM Force–Distance Curves

In this section some common and important artefacts in the acquisition of force–distance curves are discussed.

Optical Interference

As discussed in the previous section, the zero line of the approach and retraction curve is an important part of the force–distance curve since it yields the cantilever rest position, i.e. the origin of the Y axis in a force–distance curve.

In some microscopes zero lines show an oscillation [27], superimposed to the deflection signal, due to optical interference between the beam reflected by the upper side of the cantilever and that scattered by sample surface, as shown in Fig. 2.5. The laser beam (i) reaches the sample because the laser spot is larger than the cantilever surface and because the cantilever is partially transparent. The beam reflected by the sample (r_2) interferes with the beam reflected by the upper side of

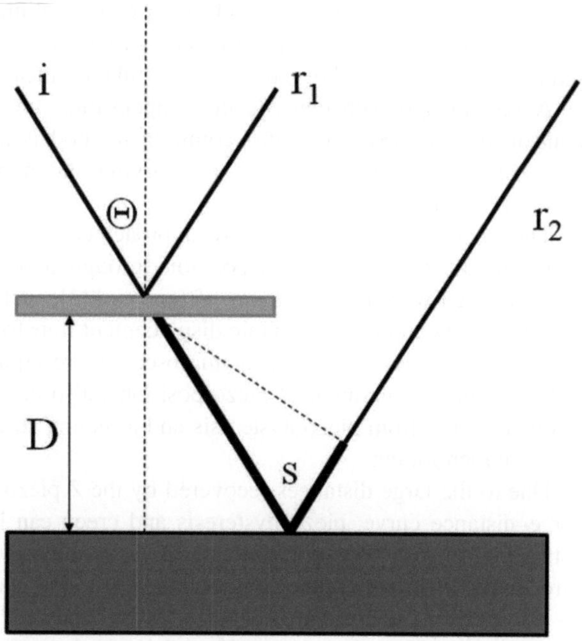

Fig. 2.5 Schematic representation of an incident laser beam (i) and of the beams reflected by the upper side of the cantilever (r_1) and by the sample (r_2). D is the cantilever-sample distance, Θ the incidence angle and s the optical path difference

the cantilever (r_1). The resulting oscillation has a spatial period determined by the laser light wavelength and by the geometry of the system.

The optical path difference s is given by [28]

$$s = n\frac{1 + \cos 2\Theta}{\cos \Theta}D, \tag{2.18}$$

in which D is the cantilever-sample distance, Θ the incidence angle and n the refractive index of the medium. If Λ is the wavelength of the laser, the spatial period of the oscillation is:

$$\Delta = \frac{\Lambda \cos \Theta}{n(1 + \cos 2\Theta)}. \tag{2.19}$$

This artefact can be corrected a posteriori by fitting the superimposed oscillation with a sinusoidal function and subtracting it. In some microscopes, such as MFP3D and Cypher (Asylum Research, Santa Barbara, CA), this artefact is eliminated by using noncoherent laser light.

Hysteresis of the Contact Lines

The scanning of the sample and/or the cantilever in X, Y and Z direction is accomplished in scanning probe microscopes through piezoelectric actuators, which provide high speed and resolution down to atomic scale. Despite their advantages, the accuracy and resolution of piezoelectric scanners are severely limited by nonlinearities, such as creep and hysteresis [29].

Artefacts due to hysteresis and creep can be eliminated through a posteriori calculations [30], but such approach cannot be used when a precise positioning is required, e.g. in nanolithography or manipulation measurements.

A second approach consists in compensating for nonlinearities by different control methods [31]; the most common method is the charge-drive technique, consisting in driving the piezo by varying the applied charge (and not the potential) [32].

The most successful and common approach consists in measuring independently and simultaneously the piezo position through a position sensor. This can be accomplished with a wide variety of sensors [33], such as capacitive or interferometric sensors and linear variable displacement transformers (LVDT).

Nowadays, most commercial microscopes are equipped with sensors for the independent measurement of piezo positions, also in *Z* direction. Nonetheless, an artefact arising from piezo hysteresis and affecting strongly force–distance curves is worth mentioning.

Due to the large distances recovered by the *Z* piezo during the acquisition of a force–distance curve, piezo hysteresis and creep can lead to a large error in the estimation of distances and affect particularly the regions of the curve where the direction is inverted, i.e. the contact line next to the maximum force.

If hysteresis and creep are not corrected or bypassed, they engender the so-called inverse path effect. In curves affected by such artefact, like the one shown in Fig. 2.6, the force in the retraction curve appears larger than the real force and eventually even larger than the corresponding force at the same displacement in the

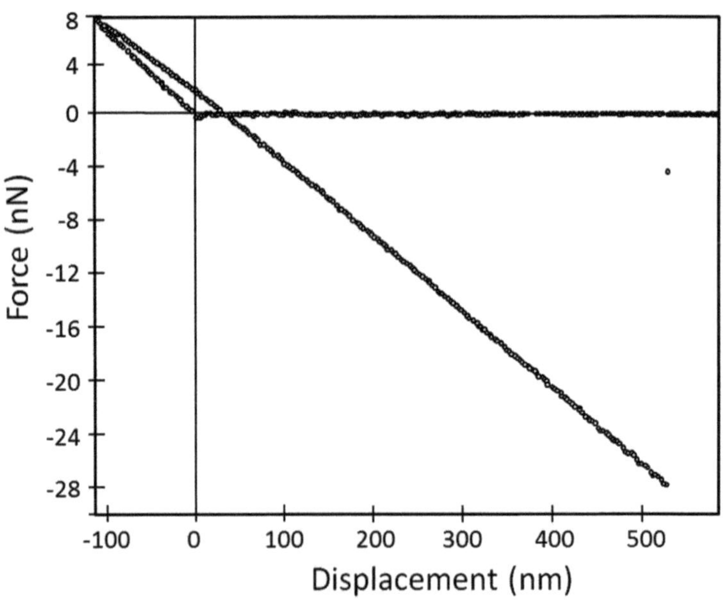

Fig. 2.6 Force–distance curve with inverse path effect

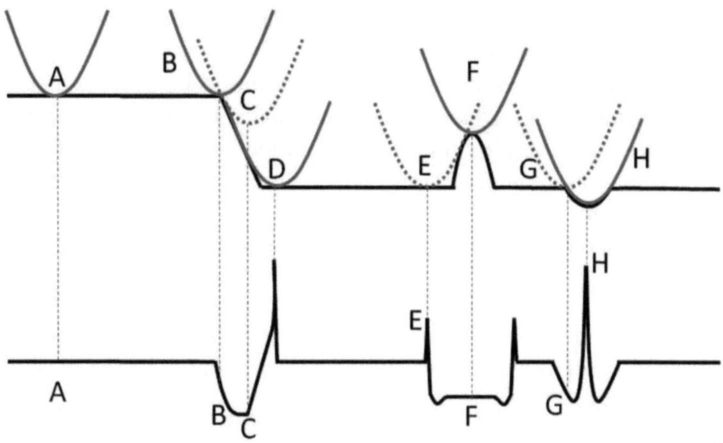

Fig. 2.7 Schematic representation of the effect of topography structures on the contact area between an AFM tip and the sample. (*Top*) Line profile of a sample with a step (B, C and D), an asperity (E and F) and a cavity (G and H). (*Bottom*) Contact length between tip and sample

approach curve. This is obviously an unphysical behaviour, since it would mean that the cantilever gains energy by deforming the sample, instead of eventually dissipating it.

Artefacts Due to Topography

One of the most important artefacts in force–distance curves measurements is due to the topography of the sample. The analysis of the deformation is performed with the assumption that the sample has a given topography, usually flat or spherical. Yet, a real sample may present steps, asperities or cavities. The effect of such topography structures on the contact area is illustrated in Fig. 2.7 (see also [34]).

Suppose that at the point A, where the sample is actually flat, the contact length at a certain load F, shown in the bottom panel of Fig. 2.7, is $2a$. Contact length is used in the following instead of contact radius, because at most of the points the sample, schematically represented by the profile in the top graph, is not symmetric with respect to the symmetry axis of the tip. This is also the reason why, at most of the points, the contact length cannot be expressed in terms of the force F. Next to the edge of the step (point B), the contact length is smaller than $2a$. At the point C, the contact length starts increasing, since also the shaft of the tip contacts the sample. The contact length reaches a maximum at point D, where both apex and shaft are in contact with the sample. This depends on the angle of both step and shaft; in the schematic representation in Fig. 2.7, step and shaft have the same angle, and the tip touches the whole step, leading to a large increase of the contact length. A similar situation occurs at the borders of an asperity (point E). On top of the asperity (point F), the contact length decreases. A decrease occurs also at the border of a cavity (point G). In this schematic representation, the cavity fits exactly the tip; hence the contact length increases considerably at the centre of the cavity (point H).

The effect on adhesion of topography irregularities such as steps and spherical caps protruding out of the surface has been calculated by Stifter et al. [35]. Due to the asymmetry of the sample, further considerations for the effect of topography structures on the measurement of Young's modulus can be done only for the top of a spherical or parabolic asperity (point F).

If the elastic modulus of the sample is E, the deformation on point F, calculated through Hertz theory, is $D = \left(\frac{F}{E_{\text{tot}}\sqrt{R^*}}\right)^{2/3}$, with $R^* = \frac{RR_{\text{asp}}}{R+R_{\text{asp}}}$, where R and R_{asp} are the curvature radius of the tip and of the asperity. If the sample is thought to be a flat plane, changes in deformation are attributed to changes in the modulus, and $D = \left(\frac{F}{E_{\text{tot}}^{\text{asp}}\sqrt{R}}\right)^{2/3}$. Hence, the apparent modulus of the asperity is given by $E_{\text{tot}}^{\text{asp}} = E\sqrt{\frac{R^*}{R}} = E\sqrt{\frac{R_{\text{asp}}}{R+R_{\text{asp}}}}$. For a tip of radius $R = 15$ nm and an asperity of radius $R_{\text{asp}} = 5$ nm, the error in the determination of the modulus is 50 %.

For forces at which the deformation of the asperity is comparable with its height, the geometry of the sample can be indeed approximated with a flat plane.

A sample with asperities indented by a smooth tip is analogous to a flat sample indented by a spherical tip with one or several 'minitips' on its surface. Such a system has been studied by Cohen [36]. In his calculation the deformation obtained with a smooth tip ($R = 1$ μm, $E_{\text{tot}} = 9$ GPa) is compared with that obtained with a rough tip (see Fig. 2.8).

The rough tip has the same radius; the roughness is modelled as a distribution of hemispherical minitips with a spacing of 10 nm. The radii of the minitips are 2 nm;

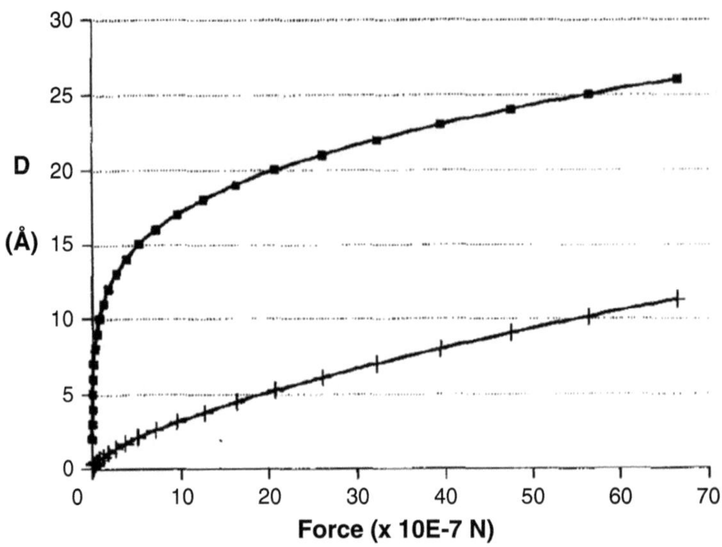

Fig. 2.8 Sample deformation obtained with a rough (*filled square*) and smooth spherical tip (*plus*). Reprinted with permission from [36]. Copyright 1992. Elsevier

their heights are represented by a Gaussian distribution centred at 2 nm and dropping to 1 % of the maximum at 0.1 and 3.9 nm.

The curves show that the initial deformation of the rough tip is considerably larger than that of the smooth tip, leading to a considerably lower apparent modulus. At about 2 nm deformation, corresponding to the mean height of the minitips, the roughness of the tip does not determine any longer the deformation, and the slopes of the deformation–force curves are the same.

The curve corresponding to the rough tip leads to a false estimation of the elastic modulus of the sample. If the curve is fitted up to a deformation of 10 Å, the resulting elastic modulus would be much lower than the real one. Moreover, if the whole curve plotted in Fig. 2.8 is considered, it could be interpreted as a deformation–force curve obtained on a mechanical double layer (see Sect. 1.10). Hence, not only the elastic modulus but also the structure of the sample would be completely misinterpreted.

Indentation Depth

An important parameter in the measurement of mechanical properties through AFM force–distance curves is the indentation depth D.

The maximum force, related to the maximum deformation or indentation, has in the praxis both an upper and a lower limit.

The upper limit is given by the tip geometry and by the sample properties. For example, if the tip apex is assumed as hemispherical, the maximum deformation should be smaller than the tip radius. The sample properties play an important role because, depending on the sample, when the maximum force is increased, plastic deformations may occur, and the system cannot be any longer modelled with an elastic continuum theory.

Several measurements have shown also the existence of a lower limit. Chizhik et al. [37] have calculated through Hertz model the elastic moduli of several polymers as a function of indentation depth. Their results are shown in Fig. 2.9.

The curves of Young's modulus as a function of indentation depth in Fig. 2.9A show a clear dependence of the modulus on the indentation depth. For $D > 20$ nm the modulus is nearly a constant for the most measured polymers (but not for the polyurethanes); for $D < 20$ nm the modulus is considerably lower (PI, PVC and PS) or larger (PU). Similar results have been obtained by Lubarsky et al. [38].

Figure 2.9B shows five curves of Young's modulus as a function of indentation depth acquired at five different locations on the PI sample surface. All curves converge to a common value (≈3 MPa) for $D > 60$ nm. Yet, the curves differ considerably from each other at small deformations. In some cases the elastic modulus is higher than 3 MPa, and the curve decreases monotonically; in some others it decreases at first and then increases again.

Dokukin and Sokolov [39] have shown that this artefact, called 'skin effect', is due to the tip-sample adhesion and to plastic deformations. The authors have determined the modulus of two different polyurethanes and polystyrene by acquiring force–distance curves with sharp and dull tips and analysing them with Oliver and Pharr, Hertz, DMT and JKR theory.

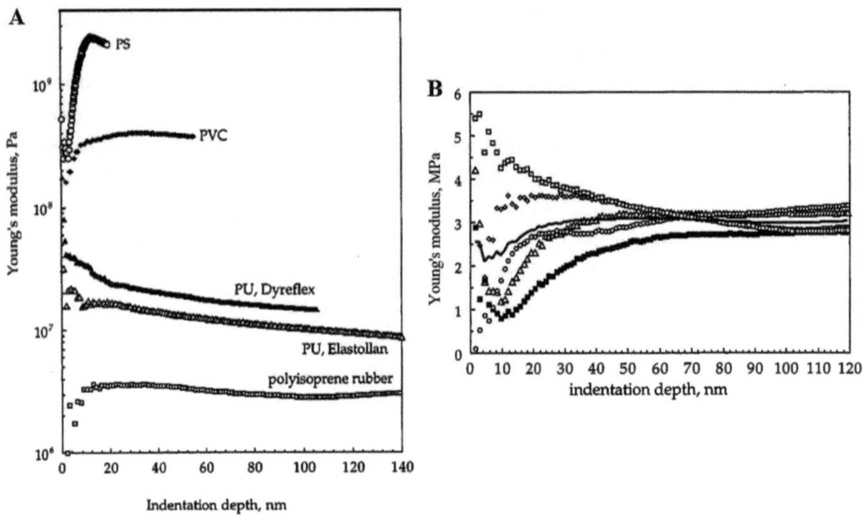

Fig. 2.9 (**A**) Elastic moduli of polyisoprene rubber (PI), two different polyurethane (PU) samples, poly(vinyl chloride) (PVC) and polystyrene (PS) as a function of indentation depth. (**B**) Young's modulus versus indentation depth D at five different locations on the PI sample surface (*markers*). Also the average profile is shown for comparison (*continuous line*). Both figures adapted with permission from [37]. Copyright 1998. American Chemical Society

With the sharp tip ($R = 22 \pm 8$ nm), a skin effect is evident in all curves plotting the calculated modulus as a function of the indentation depth, independently from the used theory. With the dull tips ($R = 810 \pm 80$ nm and 1030 ± 95 nm), the skin effect is still present in the curves obtained with Oliver and Pharr and Hertz theory, but disappears when using DMT or JKR theory. Moreover JKR theory yields a value of the modulus (2.75 ± 0.3 GPa) being in agreement with DMA measurements (2.8 ± 0.1 GPa).

The authors assess that the skin effect in the measurement with the sharp tip, engendering larger stresses, are due to the occurrence of plastic deformations. This is confirmed by the analysis of force–distance curves, showing a large hysteresis, and by inspection of the sample surface after acquisition of the curves, revealing cavities on the surface of the samples (see Sect. 1.9).

The use of a dull tip reduces the stresses engendered during the loading–unloading cycle. Nevertheless, a skin effect is still present in the values of moduli calculated through Oliver and Pharr or Hertz theory. This is due to the fact that, at small deformations, even low adhesion forces are comparable with the maximum load, and only DMT or JKR theory are suitable for modelling the deformation.

Curves on Loose Particles

Figure 2.10 shows two $D^{3/2}$ curves on an air cushion (A) and on a loose particle (B). Such curves appear similar to the curve acquired with a rough tip in Fig. 2.8. Both

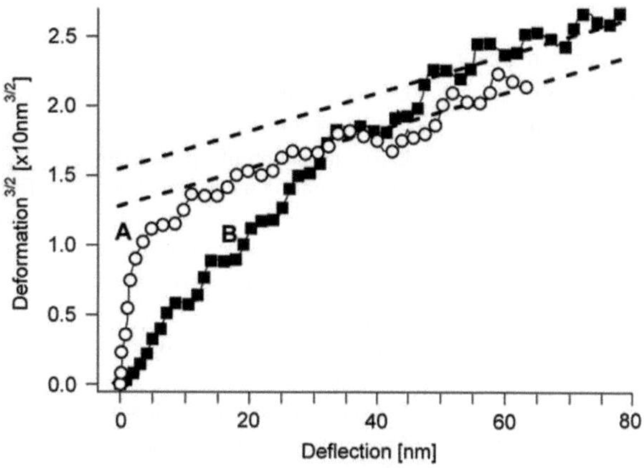

Fig. 2.10 Exemplary $D^{3/2}$-curves (*markers*) on an air cushion (**A**) and on a loose particle (**B**). The $D^{3/2}$ curve on poly(methyl methacrylate) is shown as *dashed lines*. Reprinted with permission from [40]. Copyright 2009. Elsevier

curves have been acquired on a very rough poly(methyl methacrylate) (PMMA) surface (see Sect. 4.4).

Curve A has been probably acquired on an air cushion. If a portion of the polymer sample, in a certain region, does not adhere to the rest of the film or to the substrate underneath, there is an air gap in between. At the beginning of the deformation curve, the tip pushing on the polymer compresses the air underneath. Hence, at small forces ($\delta < 5$ nm), the deformation is very large, the stiffness of the sample nearly zero, and the curve is almost a vertical line. At higher forces ($\delta > 10$ nm), when the air has been squeezed out, the $D^{3/2}$ curve has the same slope as on homogeneous PMMA, shown with a dashed line.

Curve B has been probably acquired on a loose polymer particle. Such particles can be easily shifted and deformed by the tip. Hence, the deformation at small forces ($\delta < 50$ nm) is very large, and the curve has a much larger slope than on 'normal' PMMA. Yet, the initial part of the $D^{3/2}$ curve is not almost vertical as in curve A. Eventually the tip touches the surface after deforming or shifting the particle (in this case at $\delta \approx 50$ nm), and the $D^{3/2}$ curve becomes a straight line.

2.4 Sequence of Work Steps of a Force–Distance Curves Experiment

In this section the sequence of work steps for a standard force–distance curves experiment is given in table form.

Experiment phase		Result	Related section
Cantilever calibration	Choice and mounting of a suitable cantilever		
	Characterization of tip geometry via test grating measurements	Tip shape Eventually tip radius R	Section 2.2.2
	Determination of the sensitivity via force–distance curves on stiff samples	Sensitivity Ω	Section 2.1
	Determination of the elastic constant of the cantilever	Cantilever elastic constant k_c	Section 2.2.1
Sample characterization prior to force–volume	Sample preparation and mounting		
	Measurement of the sample thickness via tapping mode images on a scratch	Sample thickness *Thin films*: Film thickness t_f	
	Characterization of sample topography	*Homogeneous samples and thin films*: Roughness *Blends*: Morphology of the blend	
	Choice of a suitable XY scan range		Section 2.3.1
	Choice of a suitable maximum force and frequency via test force–distance curves	Maximum force F_{max} Acquisition frequency ν	
Measurement	Acquisition of a force–volume		Section 2.3.1
Analysis	Automated analysis of single force–distance curves	Common origin for all force–distance curves	Section 2.3.1
	Homogeneous samples: Averaging of single curves		
	Eventually fit of the slope of approach contact line(s)	S_{eff} (k_{eff})	Section 1.2
	Calculation of deformation–force curve(s)	$D(F)$	Section 1.2
	Choice of an appropriate elastic continuum theory		Section 1.4
	Determination of the modulus	$E(W)$ *Inhomogeneous samples*: Maps of $E(W)$	Section 1.4

Wagner et al. [41] have investigated and experimentally quantified the sources of uncertainty in the measurement of Young's moduli through force–distance curves.

The authors have considered the contributions of following 11 quantities to the uncertainty of measured values of E:

1. Both measured signals, i.e. the Z-piezo voltage and the photodiode voltage.
2. Five system parameters, i.e. tip radius R, Young's modulus E_t and Poisson's ratio ν_t of the tip, Poisson's ratio of the sample ν_s and local curvature radius of the sample.
3. Four calibration parameters, i.e. Z-piezo sensitivity (conversion factor between Z-piezo voltage and elongation), sensitivity Ω, tilt angle of the cantilever and spring constant of the cantilever k_c.

The largest source of error is the sensitivity Ω. This is due to the fact that the modulus is measured from the deformation D, and D is a relatively small value calculated by subtraction of two relatively large quantities, i.e. Z (in nanometre) and δ (in nanometre). Hence, small variations of Ω have a large effect on δ and on D.

Due mainly to the contribution of Ω, the error in the determination of E for a sample with modulus $E = 10.2$ GPa is ± 8 GPa. For more compliant samples ($E \approx 1$ GPa), the error is reduced, because larger deformations are possible.

References

1. Meyer G, Amer NM (1988) Novel optical approach to atomic force microscopy. Appl Phys Lett 53:1045–1047
2. Butt H-J, Cappella B, Kappl M (2005) Force measurements with the atomic force microscope: technique, interpretation and applications. Surf Sci Rep 59:1–152
3. Neumeister JM, Ducker WA (1994) Lateral, normal, and longitudinal spring constants of atomic-force microscopy cantilevers. Rev Sci Instrum 65:2527–2531
4. Gibson CT, Watson GS, Myhra S (1996) Determination of the spring constants of probes for force microscopy/spectroscopy. Nanotechnology 7:259–262
5. Cumpson PJ, Hedley J, Zhdan P (2003) Accurate force measurement in the atomic force microscope: a microfabricated array of reference springs for easy cantilever calibration. Nanotechnology 14:918–924
6. Holbery JD, Eden VL, Sarikaya M, Fisher RM (2000) Experimental determination of scanning probe microscope cantilever spring constants utilizing a nanoindentation apparatus. Rev Sci Instrum 71:3769–3776
7. Cleveland JP, Manne S, Bocek D (1993) A nondestructive method for determining the spring constant of cantilevers for scanning force microscopy. Rev Sci Instrum 64:403–405
8. Hutter JL, Bechhoefer J (1993) Calibration of atomic-force microscope tips. Rev Sci Instrum 64:1868–1878
9. Butt H-J, Jaschke M (1995) Calculation of the thermal noise in atomic force microscopy. Nanotechnology 6:1–7
10. Clifford CA, Seah MP (2005) Quantification issues in the identification of nanoscale regions of homopolymers using modulus measurement via AFM nanoindentation. Appl Surf Sci 252:1915–1933
11. Villarrubia JS (1997) Algorithms for scanned probe microscope image simulation, surface reconstruction, and tip estimation. J Res Natl Stand Technol 102:425–454
12. Dongmo LS, Villarrubia JS, Jones SN, Renegar TB, Postek MT, Song JF (2000) Experimental test of blind tip reconstruction for scanning probe microscopy. Ultramicroscopy 85:141–153

13. Hüttl G, Beyer D, Müller E (1997) Investigation of electrical double layers on SiO_2 surfaces by means of force vs. distance measurements. Surf Interf Anal 25:543–547
14. Ducker WA, Senden TJ, Pashley RM (1991) Direct measurement of colloidal forces using an atomic force microscope. Nature 353:239–241
15. Ducker WA, Senden TJ, Pashley RM (1992) Measurement of forces in liquid using a force microscope. Langmuir 8:1831–1836
16. Butt H-J (1991) Electrostatic interaction in atomic force microscopy. Biophys J 60:777–785
17. Kappl M, Butt H-J (2002) The colloidal probe technique and its application to adhesion force measurements. Part Part Syst Charact 19:129–143
18. Gillies G, Prestidge GA, Attard P (2001) Determination of the separation in colloid probe atomic force microscopy of deformable bodies. Langmuir 17:7955–7956
19. Gillies G, Prestidge GA, Attard P (2002) An AFM study of the deformation and nanorheology of cross-linked PDMS droplets. Langmuir 18:1674–1679
20. Gillies G, Prestidge GA (2005) Colloid Probe AFM investigation of the influence of cross-linking on the interaction behavior and nano-rheology of colloidal droplets. Langmuir 21:12342–12347
21. Dimitriadis EK, Horkay F, Maresca J, Kachar B, Chadwick RS (2002) Determination of elastic moduli of thin layers of soft material using the atomic force microscope. Biophys J 82:2798–2810
22. Tan SS, Sherman RL, Ford WT (2004) Nanoscale compression of polymer microspheres by atomic force microscopy. Langmuir 20:7015–7020
23. Guo D, Li J, Xie G, Wang Y, Luo J (2014) Elastic properties of polystyrene nanospheres evaluated with atomic force microscopy: size effect and error analysis. Langmuir 30:7206–7212
24. Aimé JP, Elkaakour Z, Odin C, Bouhacina T, Michel D, Curély J, Dautant A (1994) Comments on the use of the force mode in atomic force microscopy for polymer films. J Appl Phys 76:754–762
25. Vakarelski IU, Toritani A, Nakayama M, Higashitani K (2001) Deformation and adhesion of elastomer microparticles evaluated by AFM. Langmuir 17:4739–4745
26. Buzio R, Bosca A, Krol S, Marchetto D, Valeri S, Valbusa U (2007) Deformation and adhesion of elastomer poly(dimethylsiloxane) colloidal AFM probes. Langmuir 23:9293–9302
27. Weisenhorn AL, Maivald P, Butt H-J, Hansma PK (1992) Measuring adhesion, attraction, and repulsion between surfaces in liquids with an atomic-force microscope. Phys Rev B 45:11226–11232
28. Jaschke M, Butt H-J (1995) Height calibration of optical lever atomic force microscopes by simple laser interferometry. Rev Sci Instrum 66:1258–1259
29. Bhushan B, Marti O (2004) Scanning probe microscopy – principle of operation, instrumentation and probes. In: Bhushan B (ed) Springer handbook of nanotechnology. Springer, Berlin
30. Akila J, Wadhwa SS (1995) Correction for nonlinear behavior of piezoelectric tube scanners used in scanning tunneling and atomic-force microscopy. Rev Sci Instrum 66:2517–2519
31. Butterworth JA, Pao LY, Abramovitch DY (2009) A comparison of control architectures for atomic force microscopes. Asian J Control 11:175–181
32. Fleming AJ, Leang KK (2008) Charge drives for scanning probe microscope positioning stages. Ultramicroscopy 108:1551–1557
33. Fleming AJ (2013) A review of nanometer resolution position sensors: operation and performance. Sens Actuators A: Phys 190:106–126
34. Sasaki M, Hane K, Okuma S, Torii A (1994) Scanning force microscope technique for adhesion distribution measurement. J Vac Sci Technol B 13:350–354
35. Stifter T, Weilandt E, Marti O, Hild S (1998) Influence of the topography on adhesion measured by SFM. Appl Phys A 66:S597–S605
36. Cohen SR (1992) An evaluation of the use of the atomic force microscope for studies in nanomechanics. Ultramicroscopy 42–44:66–72

37. Chizhik SA, Huang Z, Gorbunov VV, Myshkin NK, Tsukruk VV (1998) Micromechanical properties of elastic polymeric materials as probed by scanning force microscopy. Langmuir 14:2606–2609
38. Lubarsky GV, Davidson MR, Bradley RH (2004) Elastic modulus, oxidation depth and adhesion force of surface modified polystyrene studied by AFM and XPS. Surf Sci 558:135–144
39. Dokukin ME, Sokolov I (2012) On the measurements of rigidity modulus of soft materials in nanoindentation experiments at small depths. Macromolecules 45:4277–4288
40. Silbernagl D, Cappella B (2009) Reconstruction of a hidden topography by single AFM force–distance curves. Surf Sci 603:2363–2369
41. Wagner R, Moon R, Pratt J, Shaw G, Raman A (2011) Uncertainty quantification in nanomechanical measurements using the atomic force microscope. Nanotechnology 22:455703

Part II

Case Studies: Mechanical Properties of Homogeneous Polymer Films, Thin Polymer Films and Polymer Blends

Homogeneous Polymer Films 3

Abstract

This chapter presents studies on the mechanical properties of homogeneous polymer samples grouped in four cathegories: basic experiments about homogeneous polymers, measurements with colloidal probes, characterisation of viscoelastic behaviour and determination of the temperature dependence of elastic moduli.

Since atomic force microscopy is a relatively recent technique for the measurement of Young's moduli, AFM results are compared with those obtained through other techniques (instrumented nanoindentation, dynamic mechanical analysis, etc.) whenever possible.

A critical review of the results obtained, of their inherent uncertainties, mostly due to the shape of the tip and to the nanoscale topography of the sample, and of the necessary experimental effort reveals quite soon that atomic force microscopy has severe drawbacks, when compared with other well-established techniques.

Such drawbacks are mainly the fact that AFM measurements are very time-consuming and that results are affected by large uncertainties. As a matter of fact, the research field in which atomic force microscopy can be employed advantageously is the study of nonhomogeneous samples. In this case, AFM is often the only technique which can be employed.

Yet, the determination of Young's moduli of homogeneous samples is important, because it proves that AFM indentation is a feasible technique, able to yield quantitative results.

The term "homogeneous sample" needs some explanation, when used in connection with AFM. In fact, the AFM tip probes always the polymer at the interface with air or liquid. Hence, the sample can be considered homogeneous only when the mechanical properties of the volume underneath the surface are the same as the bulk properties.

B. Cappella, *Mechanical Properties of Polymers Measured through AFM Force-Distance Curves*, Springer Laboratory, DOI 10.1007/978-3-319-29459-9_3

Furthermore, it must be ensured that the thickness of the sample is much larger than the maximum indentation; otherwise, the substrate cannot be ignored and the sample is rather a mechanical double layer (see Sects. 1.10 and 4.1–4.5).

Another important issue is the range of measurable Young's moduli. This range has an upper and a lower limit.

For the measurement of large Young's moduli, say larger than 10 GPa, a very stiff cantilever is needed. This is due to the fact that the elastic modulus of the sample is determined by fitting a deformation–force curve. Considering Hertz law, the force needed to produce a deformation D is given by $F = \left(DE_{\text{tot}}\sqrt{R}\right)^{3/2}$. With $E_{\text{tot}} = 15$ GPa and $R = 10$ nm, a force of ca. 60 μN is necessary to produce a deformation of 1 nm. If the cantilever has an elastic constant $k_c = 100$ N/m, the deflection of the cantilever should be 600 nm. Depending on the microscope and on the experimental setup, such a deflection might be already in the nonlinear region of the optical lever detection system. Hence, the upper limit in the measurement of elastic moduli is determined primarily by the elastic constant of the cantilever. By employing very stiff cantilevers, with k_c considerably higher than 100 N/m, moduli of some tens of GPa could be measured. Yet, such stiff cantilevers, due to their small deflections, present many other problems: they have a very poor signal-to-noise ratio, their elastic constant is difficult to determine, and finally they cannot be used for scanning purposes since, in contact mode, they are likely to damage the sample and, in tapping mode, their oscillation is too small. Hence, such cantilevers are usually not commercially available.

When measuring very low elastic moduli in the range of 1 MPa, problems are not caused by the elastic constant of the cantilever. In this case, too large deformations must be avoided. With $E_{\text{tot}} = 1$ MPa and $R = 10$ nm, already a force of 1 nN produces a deformation of 10 nm, i.e. comparable with R. Yet, cantilevers with $k_c < 0.1$ N/m, i.e. with a deflection larger than 10 nm at $F = 1$ nN, are commercially available.

The problem by the study of very compliant materials is rather due to the fact that such materials usually exhibit a very large adhesion to the tip and form a rather high neck during retraction. In turn, due to adhesion, uncertainties arise from the choice of a suitable elastic continuum theory and from the contribution of meniscus force to the pull-off force (see Sect. 2.3.1).

In order to eliminate the capillary force and to reduce adhesion, some measurements on compliant polymers have been performed in liquid environment [1–3] or in a controlled atmosphere with low relative humidity and probes silanised with hydrophobic coatings [4]. Since such measurements, with only one exception [4], have been performed with instrumented nanoindentation and not with an AFM, they are not reported as hands-on examples.

Even in liquid environment, the adhesion force may be too large, and adding a surfactant to water may be necessary [2]. Furthermore, due to the double-layer force, it may be difficult to determine the point of contact and the origin of the deformation curve.

Further work is surely necessary, most of all with AFM, in order to measure elastic moduli smaller than 1 MPa.

3.1 Determination of the Elastic Modulus of Homogeneous Polymer Samples

The first group of experiments encompasses four hands-on examples (3.2–3.5).

In the first one [5], basic measurements of Young's moduli of polycarbonate and poly(methyl methacrylate) through force distance curves are illustrated and compared with instrumented nanoindentation.

In the two following examples, polymers with varying mechanical properties are studied. In the second hands-on example [6], the variation of the elastic modulus of polybutadiene with exposure time to air is surveyed. In the third hands-on example [7], the lateral variation of plasma-treated polystyrene is characterised; this example illustrates the possibilities of a qualitative characterisation of mechanical properties.

Finally, the last hands-on example [8, 9] reports two works dealing with Oliver and Pharr theory employed for the interpretation of measurements on numerous polymers.

Table 3.1 shows the elastic moduli of some common polymers measured with AFM, not only in the works reported in the hands-on examples. This table is not intended to be a complete list of AFM measurements of Young's moduli. Rather, only results of measurements, which can be compared with values yielded by other techniques, are reported.

3.2 Hands-on Example 1: Elastic Modulus of Poly(methyl methacrylate) and Polycarbonate

In this example a work of the groups of Cappella and Griepentrog [5] is reported.

Sample Preparation and Instrumentation For the measurement of the elastic modulus of poly(methyl methacrylate) (PMMA) and polycarbonate (PC), commercially available discs of PMMA and PC were purchased from BASF (Ludwigshafen, Germany). The samples were thicker than 4 mm and could be assumed as homogenous and isotropic. The discs were cut into smaller squares in order to fit onto the sample disc of the microscope.

AFM measurements were performed with a MFP3D microscope (Asylum Research, Santa Barbara, CA). The microscope was equipped for both measurements with a Pointprobe NCL cantilever (Nanosensor, Wetzlar-Blankenfeld, Germany) with a spring constant $k_c = 43$ N/m for the measurement on PMMA and $k_c = 39$ N/m for the measurement on PC. In both cases, the cantilever elastic constant was determined with the method of Hutter and Bechhoefer [24]

Table 3.1 Values of elastic moduli of some common polymers measured with AFM, compared with values obtained through other techniques

	Elastic modulus E (GPa)			
	Measured with AFM		Measured with other methods	
Polymer		References		References
HDPE	$0.73-0.75$	[9]	1 $0.7-1.4$	[10] [11]
LDPE	$0.16-0.19$ 0.2 0.34 ± 0.08	[9] [8] [12]	0.2 $0.14-0.3$ 0.24 ± 0.001	[10] [11] [12]
PnBMA				
Spin-coated	3.8 3.14 3.4 3.1 ± 0.3	[13] [13] [13] [14]		
Molten, at 20 °C	1 ± 0.15 0.95	[15] [16]	1.05	[15]
PC	2.45 ± 0.2 2.27 2.2 ± 0.1 2.35 2.49 ± 0.26	[5] [9] [17] [8] [12]	2.45 $2.1-2.4$ 2.68 ± 0.001	[5] [11] [12]
PET	3.78	[8]	3	[11]
PMMA	3.56 ± 0.35 3.5 $3.83-3.97$ 4.06 ± 0.59 2.8 ± 0.6 3.88 ± 0.63	[5] [18] [9] [8] [4] [12]	3.56 $2.5-3.3$ 3.2 ± 0.6 3.7 ± 0.2	[5] [11] [4] [12]
PS	3 3.37 ± 0.52 $3.56-3.72$ $3.2-3.8$ 2.9 ± 0.4 4 ± 1 3.24 ± 0.29 2.75 ± 0.3	[7] [19] [9] [20] [21] [22] [12] [23]	3.18 3.5 $2.4-3.2$ 3.24 ± 0.03 2.87 ± 0.08	[7] [20] [11] [12] [23]
PVC	1.58 1.4 1 ± 0.2	[9] [9] [22]	$1-3.5$	[11]

(see Sect. 2.2.1). On each sample, a force–volume with 10×10 curves was acquired.

Nanoindentation measurements were performed using a nanoindenter XP (Agilent Technologies, Englewood, CO) with a spherical tip.

The white light interferometer, employed for the characterisation of the nanoindenter tip, was a New-View 5022 (Zygo LOT, Middlefield, CT).

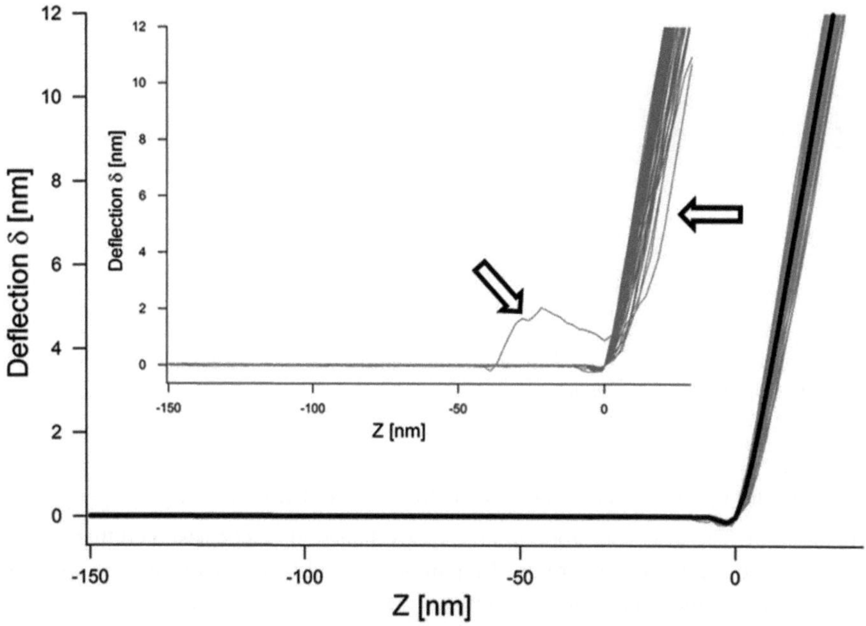

Fig. 3.1 Approach deflection–displacement curves on PMMA. The inset shows all 100 approach deflection–displacement curves. Curves affected by artefacts are highlighted with *arrows*. Such curves have been eliminated in the main graph, where the remaining 86 curves accumulate in a rather narrow cluster. The average curve is the *black thicker* one

Elastic Modulus of PMMA and PC Figure 3.1 shows a part (up to $\delta = 12$ nm) of the approach deflection–displacement curves obtained on PMMA, referred to the same origin (see Sect. 2.3.1).

The inset shows all 100 approach deflection–displacement curves.

Some curves, highlighted with arrows, deviate from the common trend and are not inside the main cluster. In particular, the curve on the left with a jump-to-contact at about -40 nm has been acquired probably on a loose polymer or dirt particle; also the curves with a significantly smaller slope of the contact line on the left part of the graphic show one or more kinks, revealing the presence of loose particles or asperities on the sample surface. These curves are affected by artefacts (see Sect. 2.3.2) and must be eliminated before the average curve is calculated.

The other 86 curves are shown in the main graph. They accumulate in a rather narrow cluster. By averaging these curves the mean deflection–displacement curve, plotted in black, can be calculated.

From the average deflection–displacement curve, the deformation D can be calculated using the definition $D = Z - \delta$, where Z is the piezo displacement and δ the cantilever deflection. The force is calculated as $F = k_c\delta$ (Hooke's law). The deformation–force curves both for PMMA and PC are plotted with circles in Fig. 3.2.

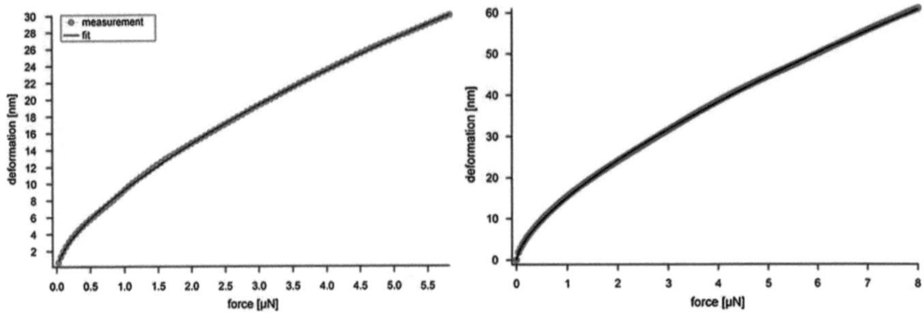

Fig. 3.2 Deformation curves obtained with AFM on PMMA (*left*) and PC (*right*). *Circles* show the experimental curves; the *black lines* are the fits with Hertz equation. Reprinted with permission from [5]. Copyright 2013. Elsevier

Both curves have been fitted with Hertz equation (Eq. 1.19), $D = \left(\frac{F}{E_{\text{tot}}\sqrt{R}}\right)^{2/3}$ (black line). The fit is very good in both cases; hence, it can be assumed that the tip apex has the shape of a sphere or of a paraboloid and Hertz theory can be applied. The elastic modulus and Poisson's ratio of the silicon nitride tip are $E_t = 245$ GPa and $\nu_t = 0.27$. With $R = 40$ nm for the PMMA measurement and $R = 25$ nm for the PC measurement (see below) and $\nu = 0.4$ for PMMA and $\nu = 0.37$ for PC, the measured elastic moduli are $E = 3.56$ GPa for PMMA and $E = 2.45$ GPa for PC.

The distribution of the curves in Fig. 3.1 can be used also to calculate the statistical error in the determination of the value of the elastic modulus. As a matter of fact, the statistical error in the determination of E is not the error in the fit, which is a measure of the quality of the fit of the average deformation–force curve, but not a measure of the variation of E in the scanned sample portion.

For every point D_i of the deformation curve, the mean value of the deformation, $\overline{D_i}$, and its standard deviation, σ_i, can be calculated. The standard deviation is used to calculate the curves $(D_{\min})_i = \overline{D_i} - \sigma_i$ and $(D_{\max})_i = \overline{D_i} + \sigma_i$. From these three curves, the mean value of the elastic modulus, $\bar{E}$, and the values $E_{\max}$ and $E_{\min}$ are obtained. The statistical error can be defined as the quantity $(E_{\max} - E_{\min})/2$.

In case of PMMA, the statistical error is 0.35 GPa (10 %); for PC, it is 0.2 GPa (8 %).

Comparison with Nanoindenter Measurements The results obtained with the AFM can be compared with nanoindenter measurements [25–27]. Such a comparison provides a confirmation of the AFM results and gives the opportunity to solve one of the most crucial problems in the quantitative measurement of mechanical properties with AFM force–distance curves, namely the determination of the tip radius.

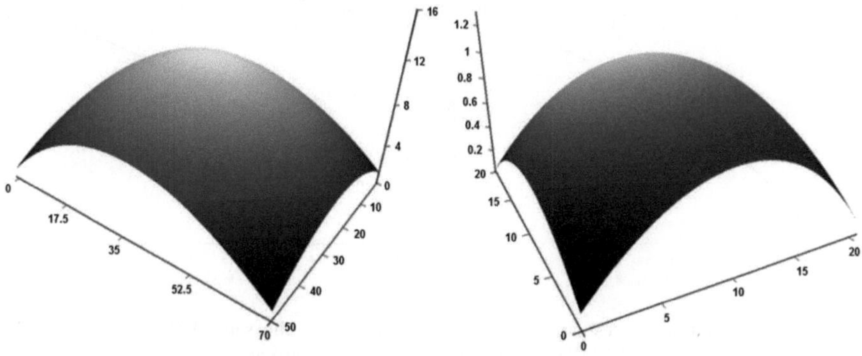

Fig. 3.3 Image of the nanoindenter tip with white light interferometer (*left*) and with AFM (*right*). Units are in microns. Adapted with permission from [5]. Copyright 2013. Elsevier

A procedure to determine the AFM tip radius through the comparison of deformation–load curves acquired by nanoindentation and AFM is described in [5] and outlined in the following.

The first step of this procedure is the determination of the shape and size of the nanoindenter tip with a white light interferometer and with an AFM.

The measurement with white light interferometer (Fig. 3.3, left) yields a radius $R = 67.6$ μm and the one with AFM (Fig. 3.3, right) $R = 68.8$ μm. The AFM measurement is quite challenging and time-consuming. The positioning of the indenter tip is crucial. Its symmetry axis has to be exactly parallel to the symmetry axis of the AFM tip, since even small deviations of some degrees would strongly affect the measured value of the radius.

The second step of the procedure is the measurement of the elastic modulus of PMMA and PC with the characterised nanoindenter tip. To this aim, the same samples used for the AFM measurement are employed.

The experimental curves (circles) shown in Fig. 3.4 have been fitted (black line) with Hertz equation. This measurement shows that Hertz theory can be applied to nanoindenter data, thus avoiding the problems associated with Oliver and Pharr theory (see Sect. 1.4.4).

The elastic modulus and Poisson's ratio of the diamond tip are $E_t = 1140$ GPa and $\nu_t = 0.07$. Using for R the mean value of both measurements, i.e. $R = 68.4$ mm, the elastic moduli result to be $E = 3.56$ GPa for PMMA and $E = 2.45$ GPa for PC.

The third step of the procedure has been anticipated in the previous section. Deformation curves are acquired on the same samples with the AFM. The values of the elastic moduli are known, since they must be the same in both measurements. As a consequence, the value of the unknown AFM tip radius can be determined as the one yielding the same value of the elastic modulus as measured with the nanoindenter.

A very effective way to compare AFM and nanoindenter deformation curves is to plot $D^{3/2} = F/(E_{\text{tot}}\sqrt{R})$ versus $F/\sqrt{R}$. In this case, the slope of the plotted lines

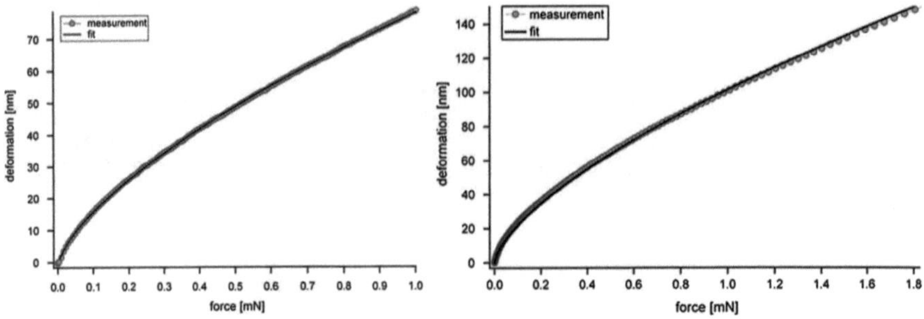

Fig. 3.4 Deformation curves obtained with nanoindenter on the same PMMA (*left*) and PC samples (*right*) as in Fig. 3.2. *Circles* show the experimental curves; the *black lines* are the fits with Hertz equation. Reprinted with permission from [5]. Copyright 2008. Elsevier

is $1/E_{\text{tot}}$ and the only difference between nanoindenter and AFM is given by the term $\left(1 - v_t^2\right)/E_t$. Since this term is much smaller than the one depending on the sample (0.3 % in case of indenter and 1.6 % in case of AFM), the plotted lines must overlap. This is shown in Fig. 3.5.

Comparing Figs. 3.2 and 3.4, it is evident that the maximum force applied by nanoindentation is about 200 times larger than that applied by AFM. In contrast, since the contact area with the nanoindenter tip is much larger, the deformation in the nanoindentation measurement is just the double of that achieved in the AFM measurement.

It is almost impossible to achieve forces of several mN with an AFM. In the experiment reported, a cantilever with $k_c = 8000$ N/m would be necessary in order to exert the same maximum load as in the nanoindentation measurement! Such stiff cantilevers are not commercially available. Secondly, even if possible, such forces would cause plastic deformation of the polymer sample, and continuum elastic theories could not be applied. Consequently, when large indentation forces are required, nanoindentation is better suited.

Large forces do not necessarily imply large pressures. The maximum pressure exerted by the indenter and by the AFM tips can be estimated as $P = F/A$, where F is the applied force and $A = \pi a^2$ is the contact area. Following Hertz theory the pressure is given by

$$P = \frac{F}{A} = F\frac{1}{\pi}\left(\frac{RF}{E_{\text{tot}}}\right)^{-2/3} = \frac{1}{\pi}\sqrt[3]{\frac{E_{\text{tot}}^2}{R^2}F}. \tag{3.1}$$

The maximum forces applied in the measurements are $F = 1$ mN for nanoindentation and $F = 5.5$ μN for AFM. These yield maximum pressures $P = 0.04$ GPa for nanoindentation and $P = 1.1$ GPa for AFM. Thus, due to the smaller radius, the pressure exerted by the AFM tip is nearly 30 times larger, even if the applied force is about 200 times smaller.

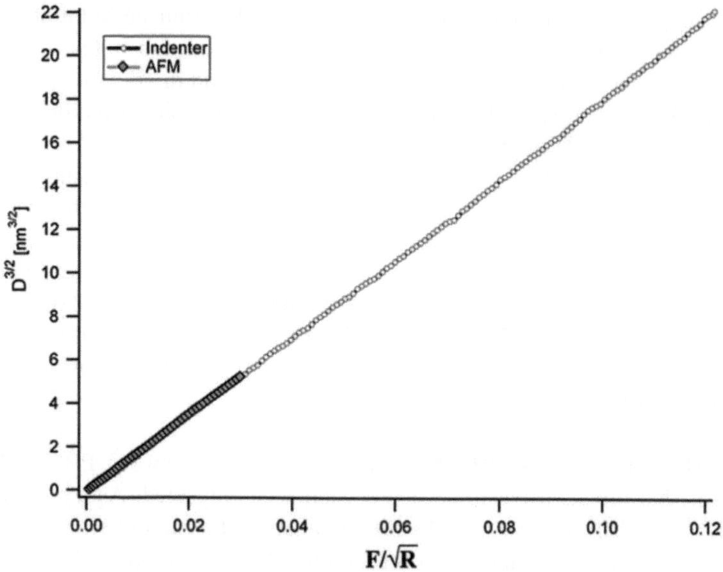

Fig. 3.5 $D^{3/2}$ vs. $F/\sqrt{R}$ curves for the indenter measurements (*grey diamonds*) and for AFM (*circles*). The two curves overlap, since the slope of the lines, $1/E_{\mathrm{tot}}$, depends mainly on the elastic modulus of the PMMA sample. Reprinted with permission from [5]. Copyright 2013. Elsevier

If only low forces are applied, the AFM is more precise than an indenter, since its vertical resolution is almost a factor of 10 higher than that of a nanoindenter. Moreover, in the deformation curve obtained through a nanoindenter, there is no equivalent of the zero line of the AFM. Consequently, the start of the indentation is difficult to detect. Usually, a certain preload is applied before the measurement is started and information about the deformation at very low forces is lost.

3.3 Hands-on Example 2: Elastic Modulus of Polybutadiene as a Function of Exposure Time to Air

In this second example, a part of the measurements exposed in [6] is outlined.

Sample Preparation and Instrumentation Polybutadiene (PB) was acquired from Sigma–Aldrich (St. Louis, MO); it contained 36 % cis, 55 % trans and 9 % 1,2 addition (all percentages are mass fractions). The molecular weight was $M_w = 420$ kDa.

The samples were prepared by using a 0.2 % PB–toluene solution for spin coating on a glass substrate at 2000 rpm for 30 s. The glass substrates were cleaned with distilled water and acetone in an ultrasonic bath and dried with nitrogen.

Before the beginning of the experiment, in order to avoid oxidation of PB, the films were stored under inert atmosphere.

Atomic force microscopy measurements have been performed using a MFP3D microscope (Asylum Research, Santa Barbara, CA) equipped with two NSC12 cantilevers (Silicon–MDT, Moscow, Russia). For the first cantilever was $k_c = 17$ N/m and $R = 40$ nm and for the second $k_c = 13$ N/m and $R = 17$ nm.

Cantilever elastic constants have been determined with the method of Hutter and Bechhoefer [24] (see Sect. 2.2.1). Tip radii were measured by recording a Tapping Mode image on a test grating with sharp tips (TGT1, NT-MDT, Moscow, Russia), as described in Sect. 2.2.2.

Elastic Modulus of Fresh PB An essential difference between PB and PMMA (or PC) is that the adhesive force of PB cannot be neglected. As a consequence, either DMT or JKR theory has to be used to analyse the curves.

Figure 3.6 shows the approach deformation–force curve recorded on a fresh homogenous PB film. The curve has been averaged out of 750 curves.

The curve has been fitted with JKR equation (black line). The choice to fit the approach curve has been discussed in Sect. 2.3.1.

The fit with JKR equation yields $W = 0.108$ J/m^2 and $E_{\text{tot}} = 6.5$ MPa. With $\nu = 0.5$, $E_t = 245$ GPa and $\nu_t = 0.27$, the elastic modulus results to be $E = 3.6 \pm 0.4$ MPa. Statistical errors have been calculated as explained in the previous example.

The curve has been fitted also with DMT equation. In this case, the fit yields $W = 0.08$ J/m^2 and $E = 3.1 \pm 0.4$ MPa.

In order to choose the better suited theory, the height of the neck formed for negative loads, $D_n \cong \sqrt[3]{RW^2/E_{\text{tot}}^2}$, is calculated in both cases. It results to be $D_n = 22$ nm for JKR and $D_n = 20$ nm for DMT theory. Since the height of the neck is much larger than the interatomic distance, JKR theory is better suited to fit deformation curves on PB and is used also in the following.

Elastic Modulus of PB as a Function of Exposure Time to Air Polybutadiene, if exposed to air, is susceptible to oxidation at room temperature due to the high number of carbon double bonds present in the molecules [28, 29]. The oxidation leads to cross-linking of the polymer chains and to an increase of Young's modulus [28, 30].

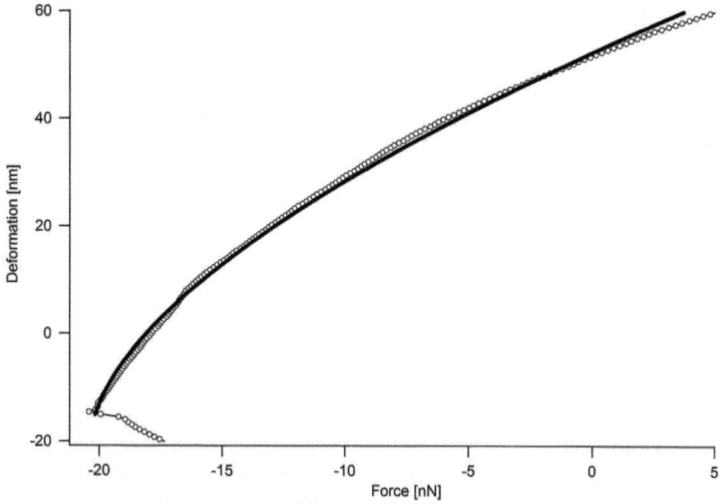

Fig. 3.6 Deformation–force curve on a fresh PB film. *Circles* show the experimental curve; the *black line* is the fit with the JKR equation

Figure 3.7 shows the increase of Young's modulus of PB with increasing exposure time to air. The first three measurements have been performed with the first cantilever and the last three with the second one.

After 360 h the elastic modulus has increased from 3.6 ± 0.4 MPa to 13.4 ± 2.2 MPa. Subsequently, the elastic modulus increases very rapidly, reaching 140 ± 15 MPa after 450 h, 525 ± 30 MPa after 480 h and 1.32 ± 0.09 GPa after 530 h. At this stage, the polymer is completely cross-linked, as the elastic modulus increases to 1.4 ± 0.15 GPa after 625 h, which is inside the error margins of the previous value. Both values are in agreement with literature values of Young's modulus of oxidised PB [30].

3.4 Hands-on Example 3: Elastic Modulus of Polystyrene Exposed to Plasma and to Toluene Vapour

In this hands-on example, the experiments in [7] are presented.

Sample Preparation and Instrumentation The samples were 15×15 mm^2 large pieces of a commercial extruded polystyrene (PS) wafer (thickness 1.2 mm, density 1.01 g/cm^3) (Goodfellow Ltd, Cambridge, UK).

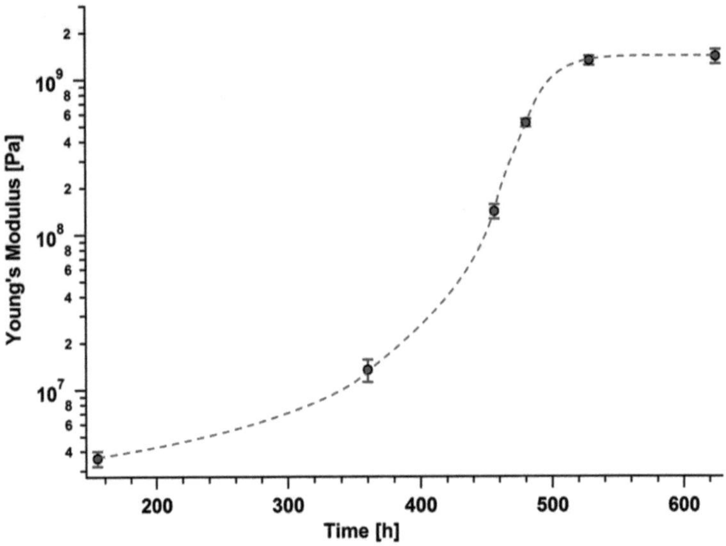

Fig. 3.7 Change of the Young's modulus of PB with increasing exposure time to air. The *dashed curve* is only a guide for the eye

The molar mass of the extruded PS was broadly distributed (mean number $M_n = 109$ kg/mol, mean weight $M_w = 284$ kg/mol) with a polydispersity index $M_w/M_n = 2.6$.

Samples were cleaned in methanol in an ultrasonic bath for two minutes. All employed solvents with ultrapure grade were purchased from Fisher Chemicals (Loughborough, UK).

A TEM grid (Plano, Wetzlar, Germany) with $45 \times 45\ \mu m^2$ large quadratic holes at a distance of 20 μm was placed as a shadow mask on the polymer sample. Afterwards, the sample was exposed for 4 min to air plasma with a power of 60 W and a pressure of 0.1 mbar in a "Femto" plasma generator (Diener Electronic GmbH, Nagold, Germany).

After plasma treatment, the grid was removed, and the sample was placed in a closed vessel with saturated toluene vapour for 2 min. After removal from the vessel, the sample was dried for 10 min and then characterised with the AFM (topography and stiffness). The exposure to vapour and the drying procedure was repeated a second time with the same sample.

The topography of the sample was acquired after the plasma treatment and after each exposure to vapour with a MFP3D atomic force microscope (Asylum Research, Santa Barbara, CA) in Tapping Mode. Pointprobe NCL cantilevers (Nanosensors, Wetzlar-Blankenfeld, Germany) with elastic constant $k_c = 40$ N/m were used. The elastic constant of the cantilever was determined with the method of Hutter and Bechhoefer [24] (see Sect. 2.2.1). The tip radius was estimated to be $R = 20$ nm.

Force–volume measurements were performed with the same cantilever.

Elastic Modulus of PS Exposed to Plasma and to Toluene Vapour Prior to modifications of the sample via plasma and vapour exposure, the elastic modulus of the original polystyrene sample was determined through a Hertz fit of the deformation curves, as shown in the previous examples.

A value of 3 GPa was obtained, in accordance with a stress–strain measurement ($E = 3.18$ GPa) and with literature values [19, 20].

Exposure of polystyrene to gas plasma leads to etching of the polymer [31], the etching rate depending primarily on the plasma power, the process time and the type of gas used. Due to the mask, parts of the sample are not exposed to the plasma.

Figure 3.8 shows the topography of the PS sample after treatment with air plasma. Covered areas, i.e. areas of the surface which were shadowed by the mask during the plasma treatment, are around 20 nm higher (brighter in the image) than the uncovered ones.

The covered areas of the sample protrude even more after exposure to solvent vapour, by about 0.4 μm after 2 min exposure and by about 1.6 μm after 4 min exposure. The covered areas are still protruding after the substrate has been removed from the solvent vapour, and even after the swollen substrate has been put into vacuum for several hours.

The height difference between covered and uncovered areas after solvent exposure can be explained by the hypothesis that the uncovered areas have been superficially cross-linked by the plasma, down to a depth depending on the plasma treatment parameters. In the covered areas, solvent molecules penetrate the polymer layer and polymer chains swell and protrude; also in the uncovered areas solvent molecules can penetrate the cross-linked superficial layer and the polymer chains underneath can swell, yet the cross-linked layer hinders polymer chains from protruding.

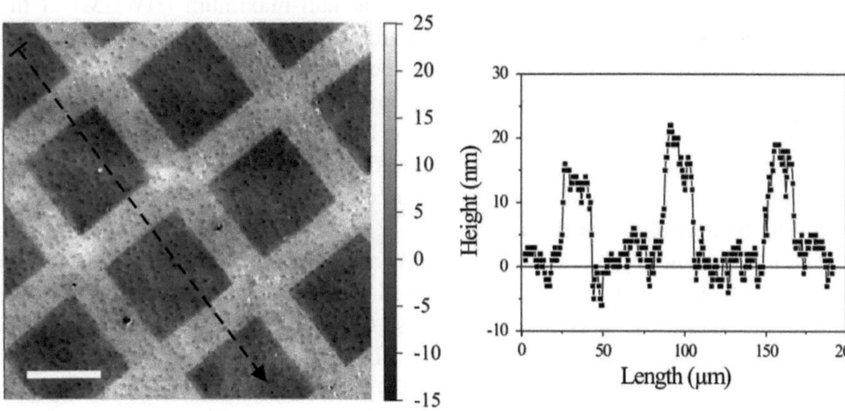

Fig. 3.8 *Left*: Topography image of a polystyrene sample exposed to air plasma through a mask. The *white scale bar* is 40 μm long. Units in the *grey scale bar* are in nanometres. *Right*: Profile taken along the *black dashed line*. Reprinted with permission from [7]. Copyright 2006. American Chemical Society

The hypothesis formulated on the existence of a cross-linked superficial layer was confirmed by characterising the mechanical properties of the polystyrene sample before and after the plasma and solvent exposures.

To examine the mechanical properties of the sample after the plasma treatment and after successive exposures to toluene vapour, force–volumes were acquired at the border between the covered and uncovered areas and the stiffness calculated from force–distance curves was mapped.

The stiffness, defined as $S_{\text{eff}} = \frac{\delta}{Z} = \frac{k_s}{k_c + k_s}$, where k_s is the elastic constant of the sample and k_c the elastic constant of the cantilever, is 1 when $k_s \gg k_c$ (see Sect. 1.2). It was calculated through a linear fit of the whole approach contact line.

The frequency of the force–distance curves (1 Hz) and the scanned area ($80 \times 80\ \mu\text{m}^2$) were the same for each measurement. The maximum cantilever deflection δ_{max} and the total number of force–distance curves were varied. Following force–volume measurements were performed:

(a) 900 curves with $\delta_{\text{max}} = 600$ nm after plasma treatment and prior to exposure to toluene vapour
(b) 900 curves with $\delta_{\text{max}} = 100$ nm after 2 min of exposure to toluene vapour
(c) 2500 curves with $\delta_{\text{max}} = 600$ nm after 2 min of exposure to toluene vapour
(d) as in measurement (c), but only 900 curves, acquired on a different area
(e) 1600 curves with $\delta_{\text{max}} = 600$ nm after 2 additional minutes of exposure to toluene vapour (total exposure time of 4 min)

The resulting histograms of S_{eff} calculated from all recorded curves are shown in Fig. 3.9.

After plasma treatment and prior to vapour exposure the stiffness histogram, (a), shows only one sharp peak at 0.997 ± 0.005. Here and in the following, the uncertainty of the stiffness is the half-width at half-maximum (HWHM) of the peak. At this stage, covered and uncovered areas have the same stiffness and their elastic constant is much higher than that of the cantilever.

After the sample has been put in vapour for 2 min, the stiffness histogram, (b), shows a broad peak centred at $S_{\text{eff}} = 0.99$. The broadening of the peak indicates that covered and uncovered areas have different stiffness, but the relatively low maximum indentation achieved in this second force–volume ($\delta_{\text{max}} = 100$ nm) does not permit to distinguish between them.

The indentation depth δ_{max} is increased in the two successive measurements to 600 nm. In the histograms (c) and (d), two well-separated peaks can be seen. Hence, with higher loads, it is possible to distinguish between the stiffness of the covered and of the uncovered areas. In particular, the stiffness of the covered areas is smaller than in the second measurement [histogram (b)], whereas the peak of the stiffness of the uncovered area is at the same position. The mean stiffness of the covered polymer is $S_{\text{eff}} = 0.968 \pm 0.003$ (c) and $S_{\text{eff}} = 0.969 \pm 0.002$ (d). The mean stiffness

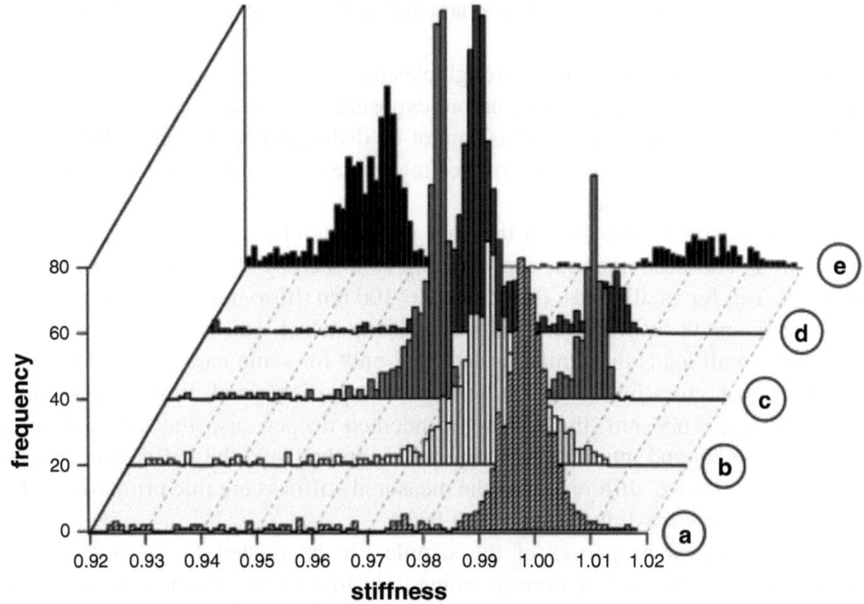

Fig. 3.9 Histograms of the stiffness S_{eff} of the PS-cantilever system after exposure to plasma and to toluene vapour for x minutes: (**a**) $\delta_{max} = 600$ nm, 900 curves, only plasma, i.e. $x = 0$; (**b**) $\delta_{max} = 100$ nm, 900 curves, $x = 2$; (**c**) $\delta_{max} = 600$ nm, 2500 curves, $x = 2$; (**d**) $\delta_{max} = 600$ nm, 900 curves, $x = 2$; (**e**) $\delta_{max} = 600$ nm, 1600 curves, $x = 4$. All histograms are normalised to 900 curves. Reprinted with permission from [7]. Copyright 2006. American Chemical Society

of the uncovered polymer is $S_{eff} = 0.997 \pm 0.004$ and $S_{eff} = 0.996 \pm 0.003$. The two measurements, performed on two different areas of the sample, confirm the reproducibility of the results obtained and show that the mechanical properties of the sample do not depend on the topography.

After an overall exposure of 4 min, still two peaks can be seen in the histogram, (e), and their separation has increased. The peak corresponding to the covered area is centred at $S_{eff} = 0.942 \pm 0.005$. The position of the higher peak is unchanged, i.e. at $S_{eff} = 1 \pm 0.004$.

The stiffness of the uncovered polymer is approximately the same in all measurements (0.996–1), including the one before exposure to toluene vapour, and is larger than the mean stiffness of the original polymer.

Measuring the stiffness of the sample, and not its elastic modulus, turns out to be a suitable strategy for the characterisation of its mechanical properties. Considering the histograms in Fig. 3.9, it is evident that the measurement of the elastic modulus would require at least two different cantilevers, a very stiff one in order to deform the polymer in the uncovered regions and a more compliant one in order to avoid plastic deformations of the polymer in the covered regions. In contrast, the measurement of the stiffness, even if it does not give information about the elastic modulus of the sample, permits to characterise the whole sample with only one cantilever.

The stiffening of the polymer through plasma treatment is a direct proof that the surface is cross-linked [32]. Also, before exposure to toluene vapour, the stiffness of the covered and uncovered areas cannot be distinguished. Hence, a thin cross-linked layer is formed also on the covered regions, even though they are not directly in contact with the plasma.

Upon exposure to the solvent the bulk polymer under the cross-linked surface layer swells and becomes much more compliant, thereby strongly influencing the stiffness. Yet, for small loads, e.g. for $\delta_{max} = 100$ nm (histogram b), the stiffness of the covered and uncovered areas cannot be distinguished. This is due to the fact that with such small loads the sample is indented only for some nanometres, and only the uppermost, cross-linked, and therefore stiff layer is probed. With higher loads, e.g. with $\delta_{max} = 600$ nm, the sample is indented deeper, also the bulk polymer, which is swollen and much more compliant, is probed, and the stiffness decreases considerably. Hence, differences in the measured stiffness are due primarily to the thickness of the cross-linked uppermost layer.

The mechanical properties of the sample can be related to its topology by comparing the stiffness map corresponding to histogram (c) and the corresponding topography. As shown in Fig. 3.10, two regions can be clearly distinguished in the stiffness map. The bright region on the right side end with $S_{eff} = 0.99 - 1$ is an uncovered area of the polymer. The grey region has a mean stiffness of about 0.97 and is about 400 nm higher than the white region, as can be seen in the right part of Fig. 3.10, where both topography and stiffness line profiles are shown. The grey region is hence a covered area.

The black border between the covered and the uncovered area is due to an artefact depending on the presence of a steep step in the topography (see Sect. 2.3.2).

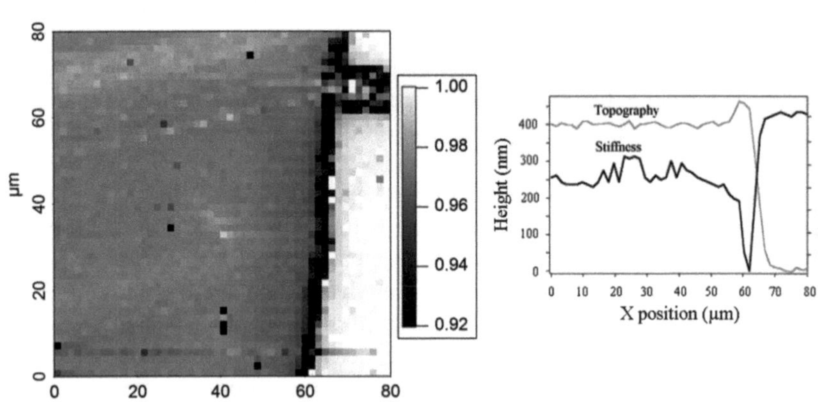

Fig. 3.10 *Left*: Stiffness map corresponding to the histogram (c). *Right*: Line profiles of the topography and the stiffness map. Reprinted with permission from [7]. Copyright 2006. American Chemical Society

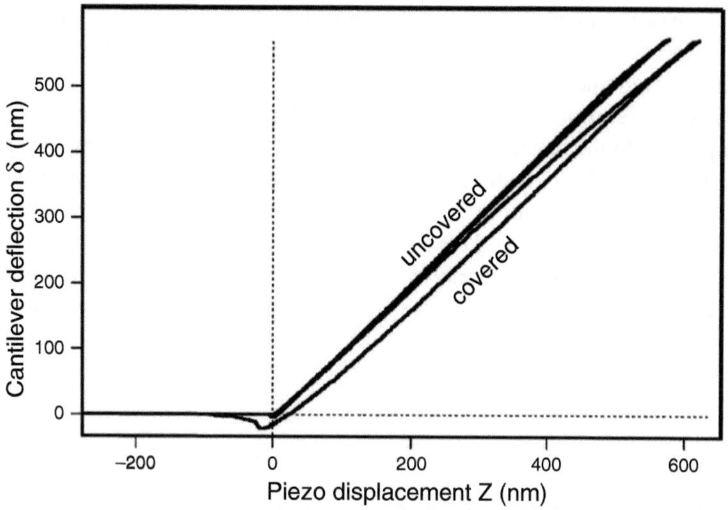

Fig. 3.11 Deflection–displacement curves within the covered and the uncovered regions of the PS surface after 4 min exposure to toluene vapour. Reprinted with permission from [7]. Copyright 2006. American Chemical Society

Considering the whole force–distance curves and not only the slope of the contact line provides a better understanding of the physical processes due to exposure to the solvent. In Fig. 3.11, two representative deflection–displacement curves acquired on the covered and uncovered polystyrene after 4 min of exposure to toluene vapour are displayed.

Following can be pointed out:

1. For deflections below about 200 nm, the two approach contact lines almost overlap.
2. At a force corresponding to about 200 nm deflection, the curve on the covered area presents a kink, after which the stiffness decreases.
3. The retraction contact line of the curve on the uncovered area overlaps with the approach contact line, i.e. there is no hysteresis. In contrast, the curve on the covered area has a non-zero hysteresis.

The overlapping of approach and retraction contact lines in the curve on the uncovered area suggests that even at indentations with $\delta_{max} = 600$ nm the cross-linked polymer behaves elastically.

For the covered area, this is true only at the beginning of the curve. The decreasing stiffness after the kink shows that the tip indents at first a rather stiff layer and then a more compliant layer underneath. This underlying layer can be plastically deformed, as indicated by the non-zero hysteresis between approach and retraction contact lines and by the presence of permanent deformations (indentation holes) left on the polymer surface after the acquisition of the curves (see Sect. 1.9).

In the measurement with $\delta_{max} = 100$ nm, the deformation is so small that no plastic deformation is achieved; as a consequence, it is not possible to distinguish between covered and uncovered areas.

These considerations are confirmed by the calculation of the energy dissipated in the sample due to the plastic deformation, defined as the area between the approach and the retraction force–distance curves above the zero axis. In measurement (b), with $\delta_{max} = 100$ nm, the dissipated energy is similar on the covered and on the uncovered areas (60 ± 20 fJ). This confirms that no plastic deformation takes place. The value of the dissipated energy on the uncovered areas is the same for all other measurements. In contrast, the dissipated energy is much larger on the covered areas both in measurements (c) (300 ± 15 fJ) and (e) (500 ± 80 fJ), showing that plastic deformations take place.

3.5 Hands-on Example 4: Comparative Analysis on the Nanoindentation of Polymers Using Oliver and Pharr Procedure

In the present hands-on example, two similar experiments are reported, that of Reynaud et al. [8] and that of Jee and Lee [9]. An additional work, where Oliver and Pharr method is used, is discussed in Sect. 6.2.

Sample Preparation and Instrumentation In the first reported article [8], four reference polymers with different known elastic moduli were chosen for calibration. The four reference polymers, purchased from Goodfellow, were low-density polyethylene (LDPE), with Young's modulus $E = 0.2$ GPa, polycarbonate (PC), with $E = 2.35$ GPa, polyethylene terephthalate (PET), with $E = 3.78$ GPa, and polyethylene naphthalate (PEN), with $E = 5.5$ GPa. The values of the elastic moduli are the nominal bulk values given by the manufacturer.

Furthermore, three samples of unknown elastic modulus were studied:

1. A biphase polymer system composed of poly(methyl methacrylate) (PMMA) nodules in a polyacrylate matrix; these two pure polymer phases have very different nominal Young's moduli, 3 GPa for PMMA and 10 MPa for polyacrylate. In order to ensure a flat surface, the blend surface was prepared by cryomicrotoming with a diamond knife at −90 °C.
2. A granule of PMMA.
3. A pellet of polyacrylate.

Force–distance curves were recorded in air with a Nanoscope III Digital Instruments microscope (Bruker, Santa Barbara, CA) at 25 °C. The silicon

Table 3.2 Polymers used in the study of Jee and Lee, together with their molecular weight (missing for PVC) and the solvent used for the sample preparation

Polymer	Acronym	M_w (kDa)	Solvent
Polyacrylic acid	PAA	240	Water
Poly(methyl methacrylate)	PMMA	120	Tetrahydrofuran
Polystyrene	PS	192	Cyclohexane
Polycaprolactam	Nylon 6	18	2,2,2-trifluoroethanol
Polycarbonate	PC	64	Tetrahydrofuran
Polyvinyl chloride	PVC		Tetrahydrofuran
Ultrahigh molecular weight polyethylene	UHMWPE	4500	Decalin
High-density polyethylene	HDPE	125	Decalin
Polyvinyl alcohol	PVA	115	6:1 mixture of DMSO and water
Low-density polyethylene	LDPE	35	Tetrahydrofuran

rectangular cantilever (Nanosensors, Neuchatel, Switzerland) used for the measurements had an elastic constant $k_c = 32 \pm 3$ N/m. The elastic constant was determined by the added end mass method (see Sect. 2.2.1).

For each polymer, force–volume measurements were performed on an area of $10 \times 10\ \mu m^2$. From each force–volume, a minimum of 10 curves were used to obtain the average indentation curves.

In the second reported article [9], ten polymers have been studied, listed in Table 3.2 together with the molecular weight M_w and the solvent used for the sample preparation. Excepted Nylon 6, which was bought from Polyscience, all other polymers were purchased from Aldrich.

The samples were prepared by compression moulding, annealing at temperatures between 110 and 200 °C depending on the polymer, and cooling to room temperature. The resulting samples were discs with a flat surface, thickness of about 1 mm and a diameter of 1.5 cm.

Nanoindentation measurements were carried out at with a XE70 AFM (Park Systems, Santa Clara, CA). The indenter was a Berkovich tip (sapphire cantilever with a diamond tip) with a spring constant of 145 N/m. The tip consists of a sharp three-sided pyramid, the base of which is an equilateral triangle with a half angle of 30°. The radius of curvature at the tip apex was nominally less than 25 nm.

Analysis of Nanoindentation with Oliver and Pharr Procedure In the work of Reynaud et al., indentation curves on the four reference polymers with known Young's moduli are fitted following the procedure of Oliver and Pharr (see Sect. 1.4.4) to obtain the parameter ε and the stiffness S. Knowing $h_{\max}$ and $P_{\max}$, the fit parameters can be factorised in the expression of the known modulus

($E_{tot} = \frac{2}{3}\sqrt{\frac{\pi}{C}\frac{S^2}{(Sh_{max} - \varepsilon P_{max})}}$, Eq. 1.46) to calculate C, the proportionality factor between the cross-sectional area of the indenter tip and the square of the contact depth, h_c^2.

Since the parameter C depends only on the tip, it can be used to calculate the modulus of PMMA and of the polyacrylate as pure samples and in the blend (hard and compliant phase, respectively).

The measured values of pure polyacrylate and of polyacrylate in the blend are 10 ± 2.4 MPa and 8.89 ± 1.74 MPa, respectively. The two values are in good agreement, but they are determined with very high uncertainty.

The measured values of pure PMMA and of PMMA in the blend are 4.06 ± 0.59 GPa and 2.54 ± 0.94 GPa, respectively. The agreement between the two experimental results is not as good as in the case of polyacrylate; furthermore, in case of PMMA, the uncertainty is even higher.

The authors claim that the discrepancy between the two values could be due to the sample preparation, since by microtoming the PMMA phase may be covered by a thin acrylate layer. Yet, errors and uncertainties due to the analysis procedure cannot be excluded.

In the work of Jee and Lee, no factor depending on the tip geometry has been calibrated. The elastic modulus is calculated from Eq. (1.42) as

$$E_{tot} = \frac{2S}{3\beta}\sqrt{\frac{\pi}{A(h_c)}}, \tag{3.2}$$

with $\beta = 1.034$ for a Berkovich indenter, $A(h_c) = 24.56\, h_c^2, h_c = h_{max} - \varepsilon P_{max}/S$ and $\varepsilon = 0.75$.

Results obtained through Oliver and Pharr analysis are compared with results yielded by Hertz theory. Yet, instead of fitting the curves, the modulus is approximated with

$$E = \frac{4(1 - \nu^2)}{3\tan\alpha}\frac{F_{max}}{D_{max}^2}, \tag{3.3}$$

with D_{max} and F_{max} the maximum deformation and maximum force, ν Poisson's ratio of the polymer and α the semi aperture of the tip (30°).

Figure 3.12 shows force–displacement curves for the ten polymers.

The non-zero areas between the approach and the retraction curves on most polymers show clearly that samples are plastically deformed. This is confirmed by topographies of the samples after indentation, showing cavities with lateral dimensions ranging from 600 nm (LDPE) to 200 nm (PAA) and depth ranging from 3.5 μm (LDPE) to 500 nm (PAA). Hence, the analysis with Hertz theory is a rather rough approximation.

The results of both analyses are listed in Table 3.3. There is a very good agreement between the two methods and the values obtained agree also with other measurements (compare Table 3.1).

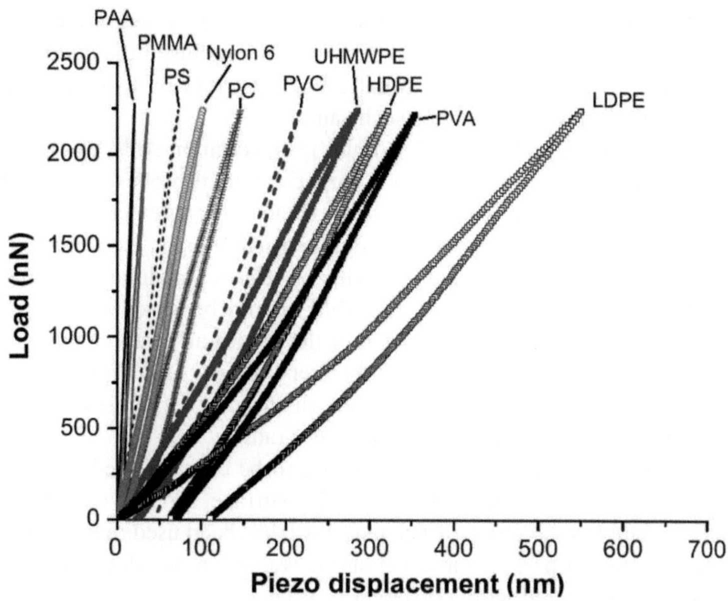

Fig. 3.12 Force–displacement curves for the ten studied polymers. Reprinted with permission from [9]. Copyright 2010. Elsevier

Table 3.3 Values of the elastic modulus *E* obtained with Oliver and Pharr analysis and with Hertz theory, together with Poisson's ratio of each polymer

Polymer	Poisson's ratio	*E* with O&P (GPa)	*E* with Hertz (GPa)
PAA	0.46	4.67	4.5
PMMA	0.35	3.83	3.97
PS	0.33	3.72	3.56
Nylon 6	0.35	3.15	3.11
PC	0.4	2.27	1.98
PVC	0.38	1.58	1.4
UHMWPE	0.46	1.39	1.13
HDPE	0.45	0.73	0.75
PVA	0.49	0.4	0.37
LDPE	0.4	0.19	0.16

3.6 Colloidal Probes

The colloidal probe technique has been described briefly in Sect. 2.2.3.

Two works with colloidal probe technique are illustrated in the following hands-on examples.

In the first one [33], colloidal probes are employed to determine the elastic modulus of silicone methacrylate microparticles. The radius of the colloidal probe, which is much larger than that of usual AFM tips, reveals to be an important advantage for the characterisation of such samples.

As shown in a work on biological samples [34], common AFM tips are sensible to the fine structure of samples, whereas larger colloidal probes tips measure the overall mechanical properties of the sample. Hence, measurements with common AFM tips are more affected by asperities and corrugations on the sample surface.

Moreover, when scanning samples with an elastic modulus of some MPa, sample deformations achieved by a usual AFM tip may be larger than the tip radius also for very small forces and the interval of the deformation–force curve that can be fitted with an elastic theory may be very short. This limitation can be circumvented employing a colloidal probe, which achieves considerably smaller deformations at the same loads, due to its larger radius.

In the second experiment [35], the polymer to be characterised is actually the colloidal probe, which is pressed onto a flat surface, in this case mica. This technique, initiated in the group of Schaefer [36], has been used mainly to measure the adhesion between particles and flat surfaces and has the drawback that each particle must be measured with a different cantilever.

3.7 Hands-on Example 5: Determination of the Elastic Modulus of Silicone Methacrylate Microparticles with Colloidal Probes

In this hands-on example, a work of Cappella et al. [33] is reported. Beyond the quantitative determination of the mechanical properties of the microparticles, the results reported show the repeatability of measurements and the possibility of employing single force–distance curves instead of averaged curves for the characterisation of samples.

Sample Preparation and Instrumentation Silicone methacrylate particles were prepared via suspension polymerisation of macromonomers in water. The silicone macromonomers were methacrylate-functionalised poly(dimethyl siloxanes) (PDMS). In order to stabilise the suspension of monomer droplets in water, partly hydrophobised SiO_2 nanoparticles were used as colloidal emulsifier. After filtration and drying, nearly spherical particles with radii of some microns were obtained, whose surface was covered with an interfacial layer of nanometre-sized SiO_2 particles.

The polymerisation of the macromonomers was initiated with two different types of radical initiators, ammonium persulfate, $(NH_4)_2S_2O_8$, and lauroyl peroxide, $[CH_3(CH_2)_{10}CO]_2O_2$. The first two letters in the nomenclature used in the work reported here are the abbreviation of the radical initiator, “AP” and “LP”. When using ammonium persulphate, the polymerisation starts at the surface and propagates into the particle core. Since the polymerisation is connected to a loss of volume, the particle contracts during the reaction. Thereby, the interface of the particle hardens at first, and the contraction leads to often observed surface wrinkles. This phenomenon does not occur when the polymerisation starts from the core, i.e. when using lauroyl peroxide. The particles prepared with lauroyl peroxide had almost ideally flat surfaces and spherical shapes.

In order to characterise particles with different cross-linking densities, the silicone macromonomers were functionalised with different ratios of methacrylate groups per siloxane unit χ, namely 0.037, 0.047, 0.055 and 0.25. The number $\chi \times 10^3$ is the second part of the names of the particles.

As last parameter, different SiO_2 particle sizes (7 and 130 nm) were used in order to characterise the influence of the interfacial layer on the mechanical properties of the particles. The size particle of the SiO_2 particle is the last part of the names of the particles.

Seven different kinds of particles were studied, grouped in two sets: AP47-7, AP55-7, AP250-7 (first set), LP37-130, LP47-130, LP55-130 and LP250-130 (second set).

Silicon wafers (100 p-type, CrysTec GmbH, Berlin, Germany) were used as substrates. The wafers were cleaned with ethanol and then coated with a thin layer of glue (Epikote 1004, Shell) heated up to 120 °C. Some droplets of the liquefied glue were blade coated to form a thin film, which was cooled down.

Spherical polymer particles, selected with an optical microscope, were placed on the wafer with help of a micromanipulator (MMO 203, Narishige, Japan) equipped with tipless cantilevers. The sample was then heated again up to 80 °C for some minutes, so that the polymer particles sank in the hot glue film, but the glue, being highly viscous, did not imbibe the particles. This was controlled by removing the polymer particles after cooling; the holes left in the glue film showed that the particles sank in the glue by a distance equal to half up to 2/3 of their radius.

Atomic force microscope (AFM) measurements were performed with two different MFP3D microscopes (Asylum Research, Santa Barbara, CA, USA). The microscopes were equipped with tipless cantilevers (TL-NCH, Nanosensors, Neuchatel, Switzerland), on which SiO_2 microspheres were glued with Epikote. The elastic constants of the cantilevers were determined by fitting the thermal noise power spectrum [24]. The radii of the colloidal probes were determined through a Tapping-Mode scanning of a TGT1 test grating (NT-MDT, Moscow, Russia) (see Sect. 2.2.2). The test grating consisted of pyramidal tips with a height of about 500 nm and a diagonal distance of 3 μm.

The type and number of particles, the properties of the cantilevers and the number of analysed force–displacement curves are listed in Table 3.4 together with the measured elastic modulus.

Table 3.4 Overview of the experimental parameters of the measurements (number of analysed particles, elastic constant of the cantilever, radius of the colloidal particle and number of curves per particle) and obtained results (elastic modulus)

Number of particles	k_c (N/m)	R_t (μm)	Number of curves	E (MPa)
AP47-7 (8 particles)				
5	35 ± 2	3.31 ± 0.13	15	2.9 ± 0.25
4[a]	36 ± 2	3.31 ± 0.13	106	3 ± 0.2
2[b]	27.4 ± 1	2.96 ± 0.13	44	2.8 ± 0.1
1	27.4 ± 2	3.44 ± 0.13	37	3.1 ± 0.1
2	27.4 ± 1	2.96 ± 0.13	47	2.9 ± 0.15
AP55-7 (7 particles)				
2	19 ± 1	3.31 ± 0.13	27	5.3 ± 0.25
2	34.4 ± 2	3.31 ± 0.13	56	5.4 ± 0.5
1	27.4 ± 2	3.31 ± 0.13	33	5.2 ± 0.25
2	27.4 ± 2	2.96 ± 0.13	58	5.3 ± 0.15
AP250-7 (7 particles)				
2	53 ± 3	3 ± 0.3	40	132 ± 10
LP37-130 (8 particles)				
3	48.5 ± 2	3.2 ± 0.13	62	2.6 ± 0.25
3	55 ± 3	3.2 ± 0.13	101	2.5 ± 0.25
2	33 ± 2	3.3 ± 0.13	80	2.2 ± 0.2
LP47-130 (10 particles)				
7	55 ± 3	3.2 ± 0.13	159	4.25 ± 0.45
3	33 ± 2	3.3 ± 0.13	65	4 ± 0.4
LP55-130 (11 particles)				
3	55 ± 3	3.2 ± 0.13	70	7 ± 0.7
2	33 ± 2	3.3 ± 0.13	157	7.6 ± 0.7
3	48.5 ± 2	3.2 ± 0.13	36	7.2 ± 0.7
3	25 ± 1	3.2 ± 0.13	62	7 ± 0.7
LP250-130 (8 particles)				
3	55 ± 3	3.2 ± 0.13	62	98 ± 15
2	33 ± 2	3.3 ± 0.13	37	97 ± 15
3	48.5 ± 2	3.2 ± 0.13	20	100 ± 15

[a]These 4 particles are a subset of the previous 5 particles characterized with the second setup and a different colloidal probe 4 months after the first measurement.

[b]These 2 particles are a subset of the previous 4 particles characterized with the first setup and a different colloidal probe 11 months after the first measurement.

Elastic Modulus of Silicone Methacrylate Microparticles The first step for the determination of the mechanical properties of the particles is the analysis of their shape. Figure 3.13 shows the topography of an AP47-7 particle, acquired in Tapping Mode with a customary AFM tip (top panel), and a profile over the apex of the particle (bottom panel). The profile was fitted with a circle function. The fact that a portion of the profile deviates from the circle function on the left side of the particle is due to the convolution with the shaft of the tip.

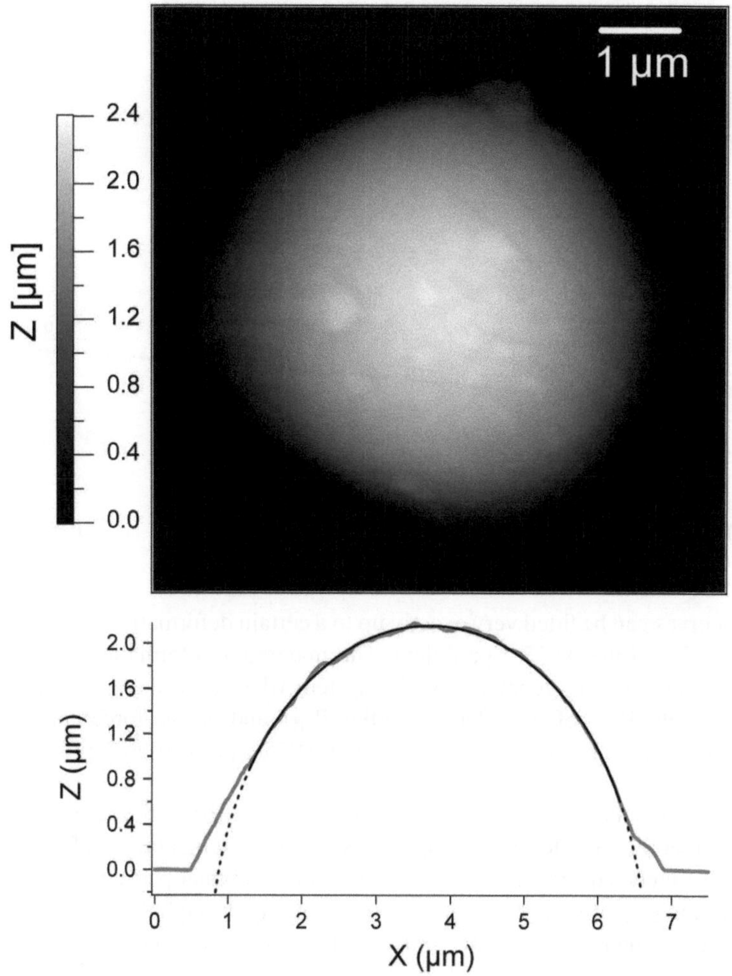

Fig. 3.13 (*Top panel*) Topography of an AP47-7 particle, acquired with an atomic force microscope in Tapping Mode. (*Bottom panel*) Profile over the apex of the particle (*grey line*), fitted with a circle function (*black line*). Reprinted with permission from [33]. Copyright 2014. Elsevier

The radii yielded by the fit of the profiles for each particle were used to calculate the effective radii $R_{\text{eff}} = RR_p/(R + R_p)$, with R and R_p the radii of colloidal probe and particle, respectively.

Figure 3.14 shows deformation curves, calculated from force–displacement curves acquired on spherical AP47-7, AP55-7 and AP250-7 particles. The deformation curves (grey circles) have been fitted with Hertz equation $D^{3/2} = F/\left(E_{\text{tot}}\sqrt{R_{\text{eff}}}\right)$ (Eq. 1.19) (black lines).

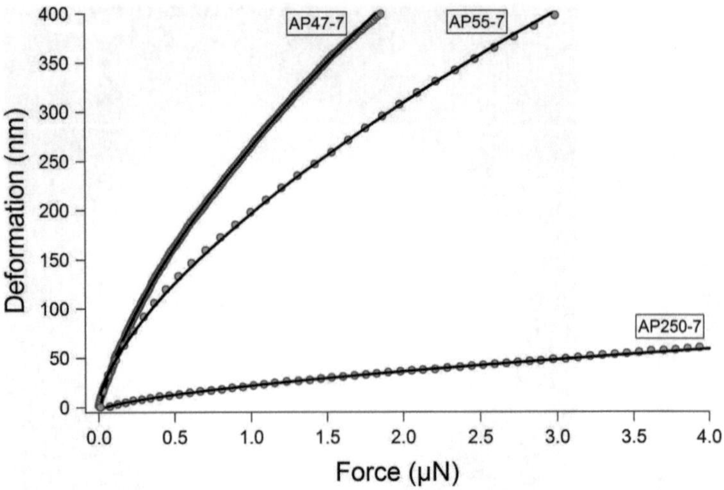

Fig. 3.14 Deformation curves acquired on spherical AP47-7, AP55-7 and AP250-7 particles (*grey circles*) and the fits with Hertz equation (*black lines*)

The curves can be fitted very exactly up to a certain deformation (e.g. 400 nm for the AP47-7 and the AP55-7 particles). The maximum deformation which can be fitted depends on the geometry of the system (the Hertz theory is valid up to approximately 10 % of the effective radius R_{eff}), and on the force applied, since very high forces (e.g. larger than 3 μN for the AP55-7 particle) may produce plastic deformations.

The good match of the experimental data with the fitting Hertz functions confirms that the particles are homogeneous, i.e. the external layer of SiO_2 particles does not influence the mechanical properties of the polymer particles. Furthermore, the adhesion is negligible, as verified experimentally through the retraction curves. This is due to the presence of the SiO_2 nanoparticles on the surface of the polymer particles.

The considered particles have radii between 1.86 and 3.32 μm (AP47-7), between 2.38 and 4.51 μm (AP55-7) and between 1.8 and 2.17 μm (AP250-7).

Knowing the elastic constant of the cantilever and the effective radius, the fit yields the elastic modulus of the polymer particles. The values of the elastic modulus for all measurements are listed in Table 3.4. In the present case, errors are calculated as the standard deviation of the values obtained from each force–deformation curve.

The measurements on particles of the same type are very consistent. Experimental data yield a mean elastic modulus $E = 2.95 \pm 0.2$ MPa for AP47-7 particles, $E = 5.3 \pm 0.3$ MPa for AP55-7 particles and $E = 132 \pm 10$ MPa for AP250-7 particles.

The very good agreement between the first three measurements on AP47-7, performed on the same particles, proves the reliability of the technique and the reproducibility of the analysis. The repeatability of the measurements, performed with different microscopes, different cantilevers and different colloidal probes over a period of 11 months, is particularly noteworthy, since the analysis is performed with single curves rather than averaged curves. Single curves are much more affected by noise and by the topography of the contact region.

For the stiffest particles, i.e. AP250-7 particles, asperities on the particle surface play an important role, since the deformation of the polymer particle and the contact radius are much smaller than for compliant particles. For example, a spherical asperity with a radius of 30 nm on the surface of an AP47-7 particle would be flattened already applying a force of some ten nanonewtons and would influence only the very beginning of the deformation curve, whereas a force of about 1.5 μN is necessary to flatten the same asperity in case of a AP250-7 particle. In this case, the asperity would influence a very large section of the curve.

Only curves acquired in a circular region around the apex of about 1 μm^2 could be fitted with Hertz equation. Outside such a region, the deformation at a certain force decreases with increasing distance from the apex of the particle. Such apparent stiffening is due to the effective increase of the contact radius when the apex of the colloidal probe contacts a point far away from the apex of the particle. In this case, the contact region between the two spheres is non-axisymmetric and the corresponding deformation cannot be fitted, due to the lack of an adequate theory. Furthermore, when force–displacement curves are acquired away from the apex, the curves show several discontinuities along the contact line, indicating that the particle is deformed also in the lateral direction and the colloidal probe slips on the polymer particle.

Deformation curves on particles of the second group (LP37-130, LP47-130, LP55-130 and LP250-130) are shown in Fig. 3.15 with the corresponding fits.

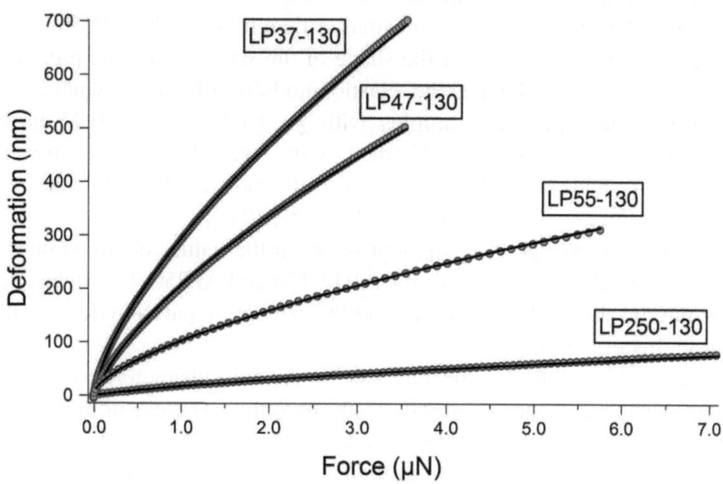

Fig. 3.15 Deformation curves acquired on spherical LP37-130, LP47-130, LP55-130 and LP250-130 particles (*grey circles*) and the fits with Hertz equation (*black lines*). Reprinted with permission from [33]. Copyright 2014. Elsevier

The radii of the particles are between 5.9 and 7.8 μm (LP37-130, whereas one particle has a radius of 20.9 μm), between 5.8 and 10.1 μm (LP47-130), between 6.1 and 14.8 μm (LP55-130) and between 8.1 and 14.7 μm (LP250-130).

Also for the second group, since all curves can be fitted very well with Hertz equation, it can be assessed that the particles are homogeneous.

Again, measurements on particles of the same type are very consistent. The mean values of the elastic moduli (see Table 3.4) are $E = 2.43 \pm 0.23$ MPa for LP37-130 particles, $E = 4.17 \pm 0.43$ MPa for LP47-130 particles, $E = 7.3 \pm 0.7$ MPa for LP55-130 particles and $E = 98 \pm 15$ MPa for LP250-130 particles.

The consistency and repeatability of the experimental values is due to the employment of a colloidal probe. Due to its dimensions, a colloidal probe is much less sensitive to the local topography of the sample. This is an important advantage, most of all for measurements on particles of the first group, which present a wrinkled surface with numerous asperities at the nanoscale.

The measured values of Young's modulus of both groups of particles correlate with the expected cross-linking density, depending on χ, the number of methacrylate groups per siloxane unit (see Sects. 3.3 and 3.4). Thereby, the modulus increases considerably between AP55-7 or LP55-130 ($E \approx 5$ MPa and $E \approx 7$ MPa respectively) and AP250-7 or LP250-130 ($E \approx 130$ MPa and $E \approx 100$ MPa respectively), i.e. between $\chi = 0.055$ and $\chi = 0.25$.

The most relevant difference between the first and the second group of particles is the thickness of the external layer of SiO_2 particles (7 nm for the first group and 130 nm for the second). Yet, no clear effect of the thickness of the SiO_2 particles layer on the mechanical properties of the polymer particles could be appraised. The moduli of the LP47-130 and LP55-130 particles are indeed slightly larger than that of the corresponding particles in the first group (AP47-7 and AP55-7), but the moduli of the LP250-130 particles are smaller than those of the AP250-7 particles.

In order to understand whether the shape of the samples and the preparation of the polymer particles influence the elastic modulus of the polymers, Young's moduli of two bulk polymer samples with $\chi = 0.047$ and $\chi = 0.25$ have been measured. The measurements, performed with a cantilever of elastic constant $k_c = 0.27$ N/m and a colloidal probe of radius $R = 500$ nm, yielded $E = 2.5 \pm 0.5$ MPa ($\chi = 0.047$) and $E = 140 \pm 25$ MPa ($\chi = 0.25$).

Since there is a very good agreement between the values obtained on the bulk samples and the values obtained for the AP47-130 and AP250-7 particles, it can be assumed that the shape of the samples and the sample preparation do not affect the elastic modulus of the polymer.

In a recent publication [37], the authors claim to have characterised effects of the size of the particles on the measured elastic modulus. The elastic modulus of PS particles, which results to be constant around 1 GPa for $R_p > 500$ nm, increases with decreasing particle radius and is larger than 3 GPa for $R_p \approx 30$ nm. After correcting the obtained values by taking into account not only the "top" deformation due to the indenting colloidal probe, but also the "bottom" deformation due to the stiff substrate, the modulus is about 2 GPa for $R_p > 500$ nm and 8 GPa for $R_p \approx 30$ nm.

The authors state that this effect can be due to increase of the free energy resulting from the confined growth, to surface effects associated with different values of the glass transition temperature at the surface of a sample and in the bulk polymer and finally to the structure of the particles, possessing a stiffer core, which influences the modulus of small particles more than that of large particles.

Yet, the hypothesis that the modulus of small particles, when measured through indentations of about 5 nm, is influenced by the substrate (see Sects. 1.10 and 4.1–4.5) is much more probable.

3.8 Hands-on Example 6: Deformation and Adhesion of Elastomer Microparticles Used as Colloidal Tips

In this hands-on example, the measurements of Vakarelski et al. [35] are reviewed. A similar experiment has been performed by Buzio et al. with poly(dimethyl siloxane) droplets [38].

Sample Preparation and Instrumentation Elastic particles, purchased from Mitsubishi Rayon Co., consist of a composite polymer with 98 % poly(diethyl hexyacrylate) and 2 % poly(acryl methacrylate) (percentages are mass fractions). The particles are coated with a few nanometres layer of poly(methyl methacrylate). Poly(diethyl hexyacrylate) is a cross-linked amorphous polymer with $T_g = -55$ °C and behaves as an elastomer of low elastic modulus above the glass transition temperature.

A freshly cleaved muscovite mica plate was used as solid flat surface.

Measurements were performed with a Multimode AFM of Digital Instruments connected to a Nanoscope III controller.

The rectangular cantilever was cut off, and a polymer particle with a radius $R_p = 1$ μm was glued on it by high-temperature melting epoxy resin. The spring constant of the cantilever, evaluated by measuring the resonance frequency of the cantilever with and without the additional mass (see Sect. 2.2), was $k_c = 2$ N/m.

In this experiment, the polymer whose properties have to be characterised is actually employed as a probe.

Elastomer Microparticles Used as Colloidal Tips The study of the mechanical behaviour of the particles starts with the characterisation of the effect of the scan rate on the load–deformation curves.

Figure 3.16a shows approach deformation–force curves at three different scan rates, $v_F = 10$, 100 and 1000 nm/s. The loading curves do not depend on the scan rate.

Unloading curves, shown in Fig. 3.16b, do depend on the scan rate. In particular, the adhesion force F_{ad} increases with decreasing scan rate, i.e. with increasing contact time between the particle and the substrate. Since the loading curves are independent of v_F, the unloading process starts from the same values δ_{max} and F_{max}.

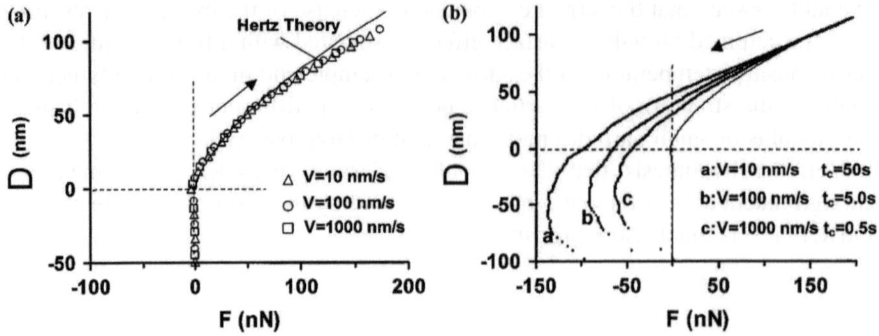

Fig. 3.16 Loading–unloading curves of a particle at $F_{\max} = 190$ nN. (**a**) Loading curves for various values of the scan rate v_F, as indicated in the legend. The *solid line* indicates the fitting curve of the Hertz theory. (**b**) Unloading curves corresponding to the curves in panel (**a**). Reprinted with permission from [35]. Copyright 2001. American Chemical Society

As can be seen in the legend of Fig. 3.16b, changing the loading rate leads to a variation of both the detachment rate v_p and the contact time t_c. In order to characterise independently the effect of these two factors, measurements have been performed by varying only one of these two parameters.

Since the adhesion force is an important factor for the application of DMT or JKR theories, the measurements at varying v_p and t_c are briefly summarised in the following, although they concern more the adhesion than the elastic modulus.

By introducing a dwell time at the maximum force, so that it is much longer than the duration of contact during the contact lines, the authors could vary v_p by keeping t_c constant. They found that the adhesion increases with increasing scanning rate. By acquiring force–distance curves at the same scanning rate, but with different dwell times, the authors found that the adhesion also increases with increasing contact time. The third factor influencing the adhesion is the maximum force $F_{\max}$. With increasing $F_{\max}$, the adhesion initially increases and then reaches a plateau.

As already illustrated in Sect. 2.3.1, the work of adhesion W turns out to be a function of both quantities: $W = W_0(t_c)\left[1 + \alpha v_p^n\right]$ (Eq. 2.17). Here, $v_p = da/dt$ is the displacement velocity of the contact line or the velocity of peeling off two surfaces, W_0 is the work of adhesion at $v_p = 0$ and α and n are material-specific constants.

Several measurement results confirm experimentally that this dependence of the adhesion force on the loading rate, although following the time–temperature superposition principle, is not related to the deformation rate of the sample. First of all, the loading curves are independent of the loading rate; second, the contact time is varied through a dwell time, i.e. when the sample is not deformed.

The authors show that, at the beginning of the retraction curve, the contact radius stays constant at the value $a_{\max}$ reached at the end of the loading process. Since the contact radius is constant, $v_p = 0$. Tip and surface start to detach first when a critical force $F_{\det}$ is reached, given by

$$F_{\text{det}} = \frac{E_{\text{tot}} a_m^3}{R} - a_m^{3/2}\sqrt{3\pi E_{\text{tot}} W_0(t_c)}. \tag{3.4}$$

For $F < F_{\text{det}}$, since $a(F) = a_m$, JKR equation (Eq. 1.28) can be written in the form:

$$D = \frac{a_m^2}{3R} + \frac{2F}{3a_m E_{\text{tot}}}. \tag{3.5}$$

Hence, the deformation is proportional to the force.

For $F > F_{\text{det}}$, the deformation can be calculated only with an iterational numerical procedure.

The important result of these measurements and of their theoretical interpretation (see also [39, 40]) is that the elastic modulus can be calculated only from the approach curves, since (1) the non-zero area between approach and retraction contact lines proves the occurring of plastic deformations as defined in Sect. 1.9, and (2) the value of F_{ad} depends on detachment velocity and on contact time.

Fitting the approach curves with Hertz equation $D = \left(\frac{F}{E_{\text{tot}}\sqrt{R_{\text{eff}}}}\right)^{2/3}$, the authors obtain for the modulus of the particles the value $E \approx 2.5$ MPa, a typical value for elastomers.

3.9 Viscoelastic Behaviour

Two experiments dealing with the characterisation of the viscoelastic behaviour of poly(n-butyl methacrylate) [15] and of polystyrene [41] are illustrated in the two following hands-on examples. These measurements are the basis for the determination of Young's moduli of both polymers outlined in Sects. 3.14 and 3.15.

Other works investigating the viscoelastic behaviour of polymer samples through models of the retraction curve and measurements of the adhesion are not discussed in hands-on examples.

In a first group of works [39, 40], the work of adhesion as a function of time is calculated through Eq. (2.17) (see Sect. 2.3.1). A part of the results of one of these works is discussed in Sect. 3.8.

In a second group of works [42–45], the creep function (Eq. 1.75) and the integration in Eq. (1.76) (see Sect. 1.8.1) are used with a given force to calculate the deformation of the sample. In these works, the interaction used for the calculation of deformations is the double-layer force (see Sect. 1.5). The double-layer force has an exponential decay and compliant samples are deformed already before contact. Hence, the determination of the distance at which deformations start is rather difficult and affects strongly the quantitative results of the experiments.

3.10 Hands-on Example 7: Viscoelastic Behaviour of Poly(*n*-butyl methacrylate)

This hands-on example reports a part of the measurements carried out in the group of Cappella [15]. The measurement of the temperature dependence of the elastic modulus is discussed in Sect. 3.14.

Sample Preparation and Instrumentation Poly(*n*-butyl methacrylate) (PnBMA) was purchased from Scientific Polymer Products (Ontario, NY). The molecular weight, the polydispersity index and the glass transition temperature given by the supplier are $M_w = 319$ kDa, $M_w/M_n = 2.58$ and $T_g = 22$ °C.

Films of PnBMA were prepared by casting saturated toluene solutions of the polymer onto a glass slide. Artefacts due to the stiff substrate were avoided by preparing very thick polymer films ($\gg$1 mm). Prior to the measurements, samples were dried at room temperature for 2 weeks.

AFM measurements were performed with a MFP3D microscope (Asylum Research, Santa Barbara, CA) equipped with a Pointprobe NCL cantilever (Nanosensors, Wetzlar-Blankenfeld, Germany) with elastic constant $k_c = 45$ N/m.

Force–distance curves were acquired at different temperatures (30, 33, 36.5, 40.5, 43.5, 46 and 51 °C). The sample was heated through a miniature metal plate controlled by a 340 temperature controller (Lake Shore Cryotronics, Westerville, OH). The temperature of the polymer surface was measured through a PT100 sensor mounted directly on the sample surface. Temperature was allowed to stabilise overnight and was constant for several days (±0.3 °C). The repeatability of the measurements was tested at the end of the temperature cycle by acquiring again curves at 30 °C.

For each temperature, force–distance curves were acquired at different frequencies (commonly 0.03, 0.1, 0.5, 1, 10 and 30 Hz). Since the minimum step of the vertical piezo elongation is 1 pm and the piezo actuator acts like a capacitor, the *Z*-ramp was assumed as continuous, and the probe rate was put equal to the acquisition frequency of the force–displacement curve.

For each temperature and frequency, 100 up to 400 force–distance curves were acquired in force–volume mode. Each force–volume measurement with 100 curves was performed on different areas of the sample (commonly 30×30 μm^2). Variations of the sample topography lead to variations of the indentation depth in a range from some 10 nm up to 500 nm.

Viscoelastic Behaviour of PnBMA Figure 3.17 shows six force–distance curves acquired on PnBMA at different temperatures and frequencies indicated in the tags.

Already at this early stage of the analysis, before calculating the deformation, some characteristics of viscoelastic behaviour can be highlighted.

Considering at first the approach curves, it is clear that they exhibit a force $F_m = k_c\delta_m$, at which the polymer goes from the instantaneous elastic regime with modulus $E_0(T, \nu)$ to the equilibrium elastic regime with modulus $E_\infty(T, \nu)$. Such a

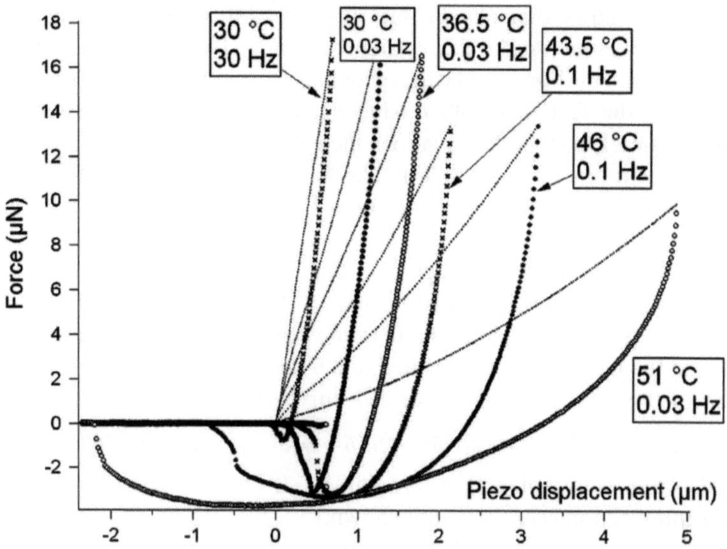

Fig. 3.17 Approach (*continuous lines*) and retraction (*markers*) force–distance curves acquired on PnBMA at different temperatures and frequencies, as indicated in the tags. Reprinted with permission from [15]. Copyright 2005. American Chemical Society

force can be seen as a kink in the approach contact line, at which the stiffness of the sample decreases.

Two effects of varying temperature and frequency can be observed in the approach curves in Fig. 3.17: the force F_m decreases with increasing temperature and/or decreasing frequency, and the stiffness of the polymer decreases with increasing temperature and/or decreasing frequency in both regions, i.e. instantaneous and equilibrium regime.

The maximum force has no effect on the approach curves. Even curves with very different F_{max} overlap, because the force F_m is not affected by F_{max}.

Going over to the retractions curves, it can be noted that the unloading curve does not overlap with the approach contact line. This is due to the fact that, during unloading, the sample does not regain immediately its shape and the force exerted by the cantilever is smaller than during the approach at each indentation depth. Thus, the viscoelastic behaviour of the polymer sample engenders a hysteresis loop. The presence of a hysteresis loop is reflected in the following two features of the curves:

1. The deformation at zero load D_p, i.e. the intercept between the retraction contact line and the axis $F = 0$; D_p would be zero for a perfectly elastic sample.
2. The dissipated energy E_D, i.e. the area between the two contact lines above the axis $F = 0$; E_D would be zero for a perfectly elastic sample.

Figure 3.17 shows that D_p, E_D and the work of adhesion W increase with increasing temperature and/or decreasing frequency. This is a consequence of the transition to the equilibrium elastic modulus, occurring at smaller and smaller forces, and of the fact that both moduli decrease with increasing temperatures and/or probe times, so that deformations increase.

It is important to emphasise that, unlike the force F_m and the stiffness, D_p, E_D, and W depend also on the maximum force $F_{\max}$. In fact, retraction curves with different $F_{\max}$ do not overlap.

Figure 3.18A shows the average $D^{3/2}$ curve calculated from a force–volume measurement with 100 curves at 30 °C and 30 Hz. Figure 3.18B shows 9 average $D^{3/2}$ curves acquired at different temperatures and frequencies. The temperatures indicated in the figure have been calculated by means of Williams–Landel–Ferry equation (see Sects. 1.7 and 3.14). All curves are fitted with the hyperbolic function $D^{3/2} = \beta\delta - \varepsilon + \sqrt{\alpha^2\delta^2 - 2\varepsilon(\beta - \gamma)\delta + \varepsilon^2}$, Eq. (1.86). In the left panel, also the asymptotes $D^{3/2} = \gamma\delta$ for $\delta \to 0$ and $D^{3/2} = (\beta + \alpha)\delta$ for $\delta \to \infty$ and their intersection $(\varepsilon/\alpha, \gamma\varepsilon/\alpha)$ are shown.

The hyperbolic function is able to reproduce the $D^{3/2}$ curves at each temperature.

Unlike a $D^{3/2}$ curve obtained on a perfectly elastic polymer, which is a straight line, the $D^{3/2}$ curves in Fig. 3.18 present two linear regions (see Sect. 1.8.3). The first region at low deflections is the instantaneous regime with modulus $E_0(T)$ and the second the equilibrium regime with modulus $E_\infty(T)$. In these two regions, the fitting hyperbola can be approximated with two straight lines $D^{3/2} = \gamma\delta$ and $D^{3/2} = (\beta + \alpha)\delta$. The slope of the curve in the equilibrium regime, $\beta + \alpha$, is always larger than the slope in the instantaneous regime, i.e. γ. The parameters $\beta + \alpha$ and γ are inversely proportional to the moduli $E_\infty(T)$ and $E_0(T)$, respectively.

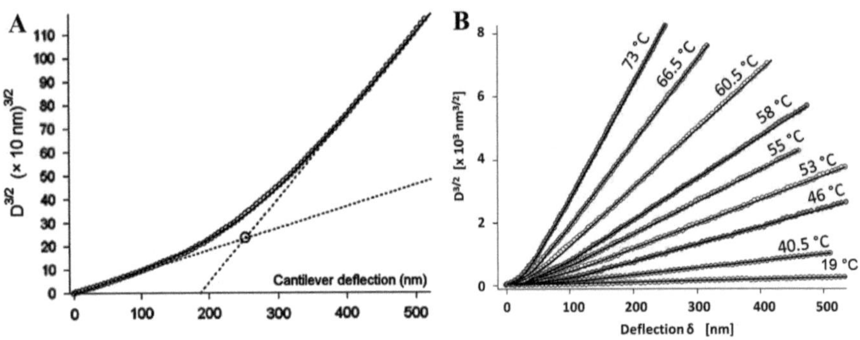

Fig. 3.18 (**A**) Fit of an average $D^{3/2}$ curve acquired at 30 °C and 30 Hz (*open circles*) with the hyperbolic function (Eq. 1.86) (*thick continuous line*). Also, the asymptotes for $D^{3/2} = \gamma\delta$ and $D^{3/2} = (\beta + \alpha)\delta$ and their intersection $(\varepsilon/\alpha, \gamma\varepsilon/\alpha)$ are shown. (**B**) $D^{3/2}$ curves (*circles*) obtained at different frequencies and temperatures, fitted with the hyperbolic function (Eq. 1.86) (*continuous line*). The temperatures indicated in the figure have been calculated by means of Williams–Landel–Ferry equation (see Sect. 3.14). The *left panel* has been adapted with permission from [15]. Copyright 2005. American Chemical Society

Hence, at each temperature, E_∞ is lower than E_0, as expected. Furthermore, as already said, both moduli decrease with increasing temperature.

Between the two linear regions, there is a transition region where the slope slowly increases. The centre of the transition region is placed at the point $(\varepsilon/\alpha, \gamma\varepsilon/\alpha)$, shown in the left panel with a circle.

Figure 3.19A shows the first derivative of six of the $D^{3/2}$ curves shown in Fig. 3.18A and Fig. 3.19B shows the second derivative of the $D^{3/2}$ curve obtained at 66.5 °C.

The first derivatives and the second derivative are fitted with the sigmoidal function $\frac{\partial D^{3/2}}{\partial \delta} = \beta + \frac{\alpha^2\delta - \varepsilon(\beta-\gamma)}{\sqrt{\alpha^2\delta^2 - 2\varepsilon(\beta-\gamma)\delta + \varepsilon^2}}$, Eq. (1.87), and with the bell function $\frac{\partial^2 D^{3/2}}{\partial \delta^2} = \frac{\varepsilon^2\left[\alpha^2 - (\beta-\gamma)^2\right]}{\sqrt{\alpha^2\delta^2 - 2\varepsilon(\beta-\gamma)\delta + \varepsilon^2}^3}$, Eq. (1.89), respectively.

For PnBMA, the transition region between instantaneous and equilibrium regime is placed at rather low forces and deformations or, in other words, occurs after rather short times. As a consequence, the instantaneous plateau cannot be seen, whereas the equilibrium plateau is reached at relatively low forces. Also, the second derivative is zero only at high forces and deflections, and not for $\delta \to 0$. Nevertheless, the maximum of the second derivative at $\delta \approx 17$ nm can be seen clearly.

The occurrence of the transition region between instantaneous and equilibrium regime at low forces can be seen quite clearly by plotting the same curves in Fig. 3.18B transformed with the coordinate transformation in Eq. (1.90), i.e. $x = \delta - \delta_m$ and $y = D^{3/2} - \beta\delta + \varepsilon$, with $\delta_m = \varepsilon(\beta-\gamma)/\alpha^2$ (see Sect. 1.8.3). As can be seen in Fig. 3.20, with exception of the curve at 19 °C, all curves are in the first quadrant, i.e. in the equilibrium regime.

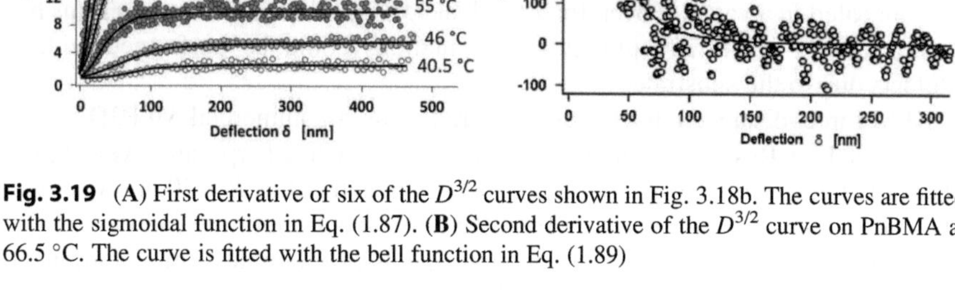

Fig. 3.19 (**A**) First derivative of six of the $D^{3/2}$ curves shown in Fig. 3.18b. The curves are fitted with the sigmoidal function in Eq. (1.87). (**B**) Second derivative of the $D^{3/2}$ curve on PnBMA at 66.5 °C. The curve is fitted with the bell function in Eq. (1.89)

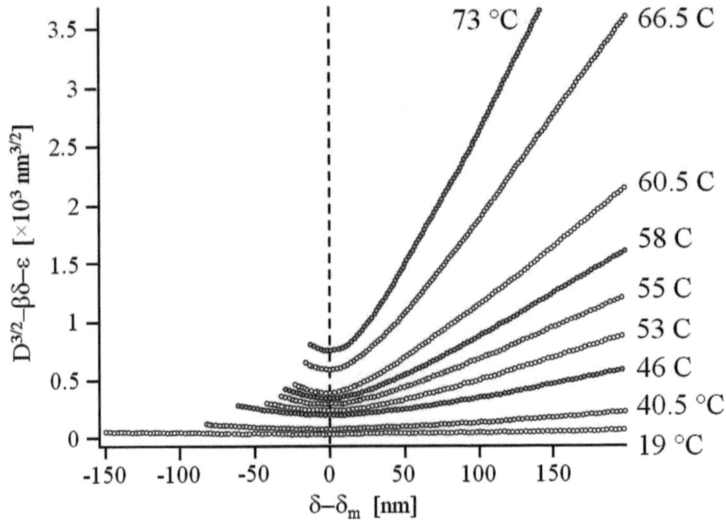

Fig. 3.20 $D^{3/2}$ curves shown in Fig. 3.18b, transformed with the coordinate transformation in Eq. (1.90)

3.11 Hands-on Example 8: Viscoelastic Behaviour of Polystyrene

This hands-on example reports the results of a work in the group of Cappella [41].

The determination of Young's modulus of PS as a function of temperature is described in Sect. 3.15.

Sample Preparation and Instrumentation Polystyrene (PS) with two different molecular weights, 4.2 kDa (PS_{4k}) and 62.5 kDa (PS_{62k}), was purchased from Polymer Standards Service (Mainz, Germany). The glass transition temperatures indicated by the provider are 57 °C and 97 °C, respectively. The polydispersity index is $M_w/M_n = 1.05$ for both polymers.

Films of both polymers were cast onto glass slides from solutions in toluene and then annealed in a vacuum oven for 1 week at 150 °C. Since the resulting films had a thickness of about 250 μm, large indentations could be performed without artefacts due to the substrate.

AFM force–distance curves were acquired using a commercial MFP3D microscope (Asylum Research, Santa Barbara, CA). An Ultrasharp cantilever (Mikro Masch, Estonia) with spring constant $k_c = 15$ N/m and radius $R \approx 20$ nm was used. The spring constant was measured from the noise spectrum of the cantilever [24].

As in the work discussed in the previous hands-on example, a miniature metal plate at the basis of the polymer was heated using a 340-temperature controller (Lake Shore Cryotronics, Westerville, OH). The temperature was allowed to equilibrate overnight before each measurement. The surface temperature was monitored using a PT100 sensor fixed onto the polymer surface and was constant (±0.3 °C) for several days.

Force–displacement curves were collected at various temperatures: 30, 42, 54, 61, 67, 75, 82 and 95 °C for PS_{4k} and 30, 41, 52, 62 and 84 °C for PS_{62k}. At each experimental temperature force–displacement curves were acquired at various frequencies (usually 30, 10, 1, 0.5, 0.1 and 0.03 Hz).

As already stated in the previous hands-on example, since the minimum step of the vertical piezo displacement is 1 pm and the piezo actuator acts like a capacitor, the displacement can be assumed as continuous, and the probe rate is actually the frequency of the force–displacement curve.

For each measurement at a certain temperature and frequency, a variable number of force–volume measurements with 100 × 100 curves were performed. Each force–volume measurement was performed on a different surface area of the films (commonly 20 × 20 μm^2). Due to variations in the sample topography, the curves had different maximum indentation. In total, more than ca. 15,000 curves have been taken into account for the analysis.

Viscoelastic Behaviour of PS Figure 3.21a shows six approach force–distance curves obtained on PS_{4k} at various temperatures and frequencies, as indicated in the figure. The retraction curves (broken lines) are shown only for the curves acquired at 30 °C and 30 Hz and at 75 °C and 0.1 Hz.

Figure 3.21b shows five force–distance curves acquired on PS_{62k} at various temperatures and frequencies, as indicated in the figure. Except the one at 52 °C and 0.1 Hz, the curves were shifted horizontally for clarity. In both graphs, double-bordered squares indicate the centres of the transition region between the instantaneous and the equilibrium regime.

By comparing these curves with the ones on PnBMA shown in the previous hands-on example, it is evident that, for PS_{62k}, the transition from the instantaneous to the equilibrium regime occurs at higher forces and almost the entire range of experimental loads is in the instantaneous regime. Also, the position of the centre of the transition region does not depend sensibly on the temperature. Moreover, the deformation at zero load D_p, the dissipated energy E_D and the work of adhesion W are considerably smaller than for PnBMA. Finally, except for the curve at the highest temperature, the stiffness of the sample does not depend considerably on the temperature.

PS_{4k} exhibits a mixed behaviour. At low temperatures, the curves resemble those acquired on PS_{62k}. At temperatures higher than 60 °C, the transition from the instantaneous to the equilibrium regime occurs at lower and lower forces with increasing temperature; furthermore, D_p, E_D and W severely increase, and the stiffness depends considerably on the temperature.

Figure 3.22 shows $D^{3/2}$ curves obtained on PS_{4k} at different temperatures and frequencies. The temperatures indicated in the figure have been calculated by means of Arrhenius and Williams–Landel–Ferry equation (see Sects. 1.7 and 3.15). The curves are fitted with the hyperbolic function

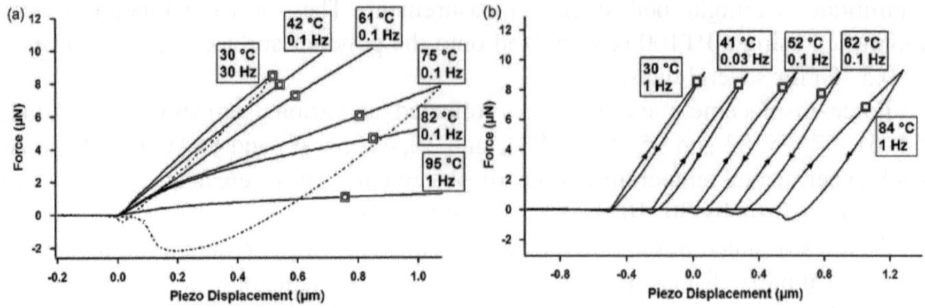

Fig. 3.21 (**a**) Approach force–distance curves obtained on PS_{4k} at various temperatures and frequencies, as indicated in the figure (*solid lines*). The retraction curves (*broken lines*) are shown only for the curves obtained at 30 °C and 30 Hz and at 75 °C and 0.1 Hz. (**b**) Force–distance curves acquired on PS_{62k} at various temperatures and frequencies, as indicated in the figure. Except the one at 52 °C and 0.1 Hz, the curves were shifted horizontally for clarity. The centres of the transition region between the instantaneous and the equilibrium regime are indicated by *double-bordered squares*. Reprinted with permission from [41]. Copyright 2005. Elsevier

$D^{3/2} = \beta\delta - \varepsilon + \sqrt{\alpha^2\delta^2 - 2\varepsilon(\beta - \gamma)\delta + \varepsilon^2}$ (Eq. 1.86). Again, the hyperbolic function is suitable to reproduce the $D^{3/2}$ curves at each temperature.

Other than on PnBMA, up to 62.5 °C, the $D^{3/2}$ curves exhibit clearly a linear region at low deflections and forces, i.e. the instantaneous linear region. Only at higher temperatures the stiffness depends sensibly on the temperature and $D^{3/2}$ is not proportional to the deflection.

Figure 3.23 shows the first derivative of five of the $D^{3/2}$ curves in Fig. 3.22. The curves are fitted with the sigmoid $\frac{\partial D^{3/2}}{\partial \delta} = \beta + \frac{\alpha^2\delta - \varepsilon(\beta-\gamma)}{\sqrt{\alpha^2\delta^2 - 2\varepsilon(\beta-\gamma)\delta + \varepsilon^2}}$ (Eq. 1.87).

Due to the fact that the transition region is located at higher forces, the instantaneous plateau $\partial D^{3/2}/\partial\delta = \gamma$, inversely proportional to $E_0(T)$, can be clearly seen in the three curves at lowest temperatures. Instead, the equilibrium plateau $\partial D^{3/2}/\partial\delta = \beta + \alpha$, inversely proportional to $E_\infty(T)$, cannot be reached, even in the curve at 84 °C, as can be seen through the fit, plotted also for forces higher than the maximum force.

The second derivative of the $D^{3/2}$ curve at 21 °C, fitted with the bell function $\frac{\partial^2 D^{3/2}}{\partial \delta^2} = \frac{\varepsilon^2\left[\alpha^2 - (\beta-\gamma)^2\right]}{\sqrt{\alpha^2\delta^2 - 2\varepsilon(\beta-\gamma)\delta + \varepsilon^2}^3}$ (Eq. 1.89), is shown in Fig. 3.24. Other than for PnBMA (Fig. 3.19B), the zero plateau can be seen at low forces and short times, confirming that the transition to the equilibrium regime is located at very large deflections, i.e. at very long times.

Figure 3.25 shows $D^{3/2}$ curves on PS_{4k} transformed through the coordinate transformation in Eq. (1.90). As already said, this way of plotting the curves shows in a rather perspicuous way whether the polymer at small forces and short times is still in the instantaneous regime or already in the equilibrium regime. In case of PS_{4k}, the applied forces are smaller than the centres of the transition regions in which the modulus of the sample goes from $E_0(T)$ to $E_\infty(T)$. Hence,

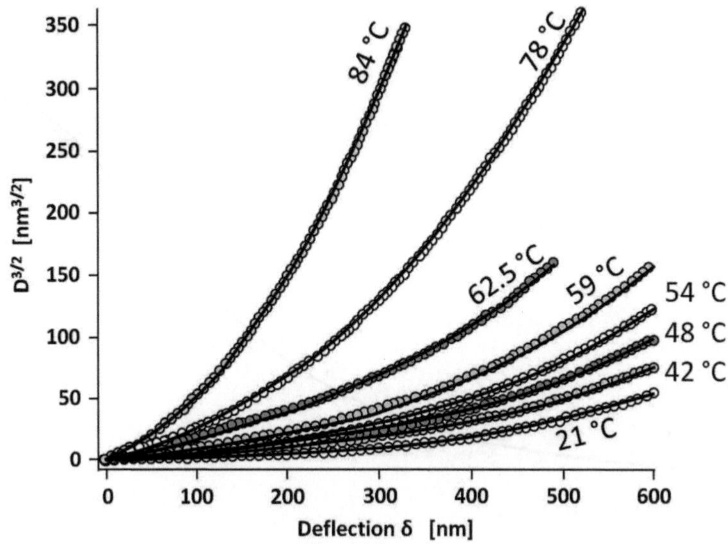

Fig. 3.22 $D^{3/2}$ curves on PS_{4k} obtained at different temperatures and frequencies. The temperatures indicated in the figure have been calculated by means of Arrhenius and Williams–Landel–Ferry equation (see Sect. 3.15). The curves are fitted with the hyperbolic function (Eq. 1.86)

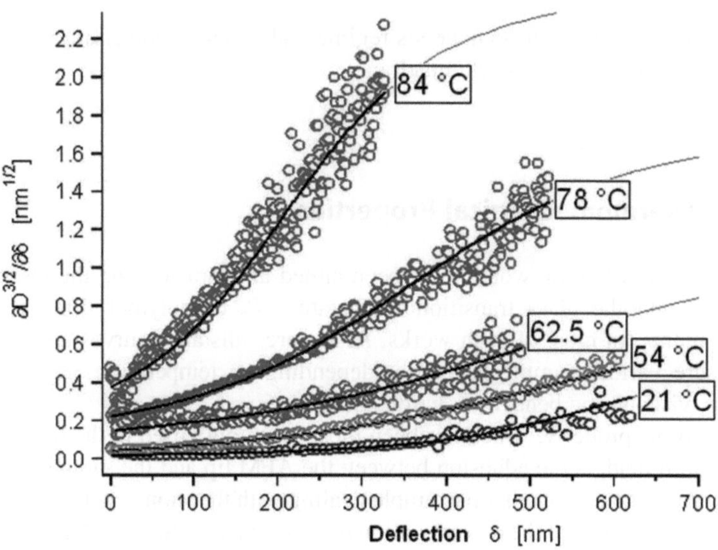

Fig. 3.23 First derivative of five of the $D^{3/2}$ curves shown in Fig. 3.22. The curves are fitted with the sigmoid in (Eq. 1.87)

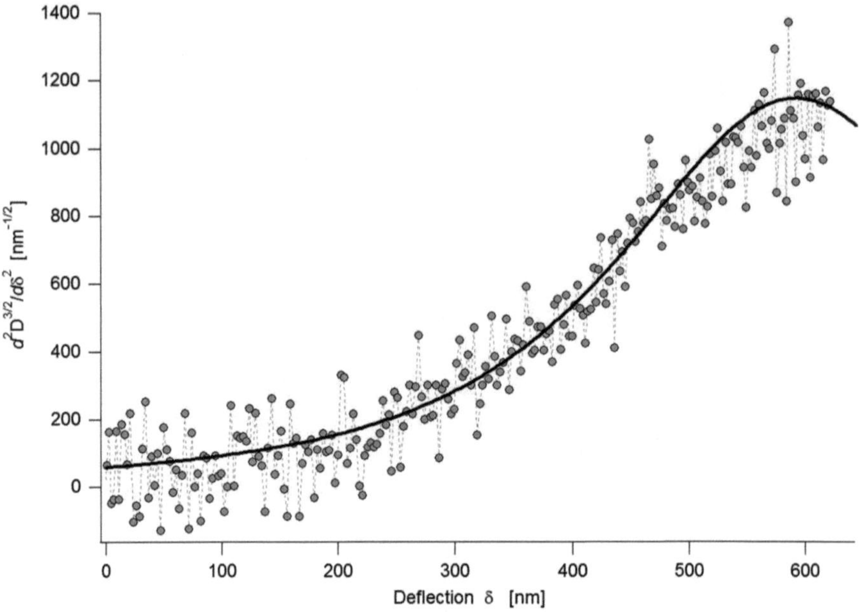

Fig. 3.24 Second derivative of the $D^{3/2}$ curve at 21 °C shown in Fig. 3.22. The curve is fitted with the bell function in Eq. (1.89)

deformations are in the instantaneous regime and, at each temperature, the curves are almost entirely in the fourth quadrant.

3.12 Thermomechanical Properties

Several recent scientific works have been aimed at characterising the viscoelastic behaviour and the glass transition temperature T_g of polymer films via AFM measurements. In most of such works, AFM force–distance curves are employed to measure various sample properties depending on temperature, e.g. adhesion, stiffness, viscoelastic behaviour, hardness, etc.

The sample property, which has been most commonly used in the first works in this research field, is the adhesion between the AFM tip and the sample.

Therefore, the first hands-on example dealing with thermomechanical properties reviews works aimed at the determination of T_g via measurements of the tip-sample adhesion and in particular reports experiments performed in the group of Tsui [46].

Later on, Cappella and Stark [47] have shown that the temperature dependence of the tip-sample adhesion of polymers is due to the temperature dependence of the stiffness and of the viscoelastic behaviour of the sample, affecting the tip-sample contact area. Their analysis, also reported in Sect. 3.13, shows that the viscoelastic

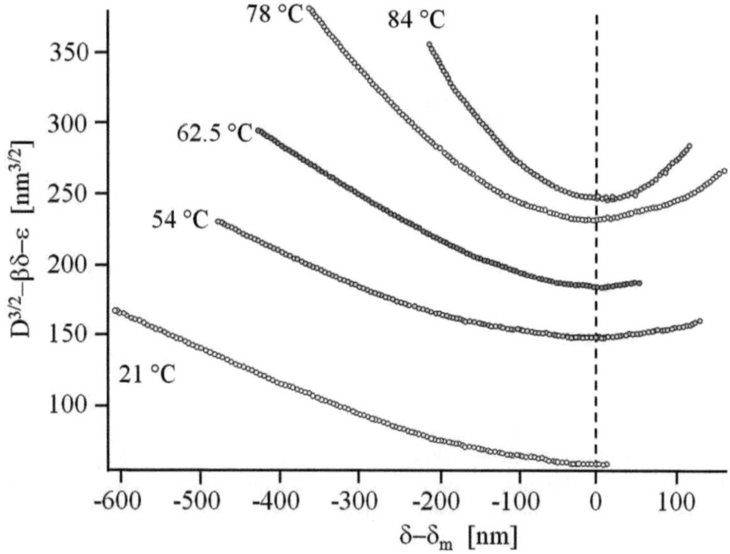

Fig. 3.25 $D^{3/2}$ curves on PS with molecular weight 4 kDa. The curves are transformed with the coordinate transformation Eq. (1.90)

behaviour of a polymer should not be studied by characterising the temperature dependence of the adhesion, but rather by describing the thermomechanical properties.

This is illustrated in the two following hands-on example, where the elastic moduli of poly(*n*-butyl methacrylate) and of two polystyrene samples are determined as a function of temperature. The characterisation of force–distance curves acquired on the same samples at different temperatures and frequencies has been discussed in Sects. 3.10 and 3.11.

3.13 Hands-on Example 9: Studying the Glass-to-Rubber Transition of Poly(*tert*-butyl acrylate) Using Adhesion Measurements

The work reported in this hands-on example has been chosen as representative of a small group of publications dealing with the determination of the glass transition temperature of polymers via measurements of the tip-sample adhesion.

Two works in this category are briefly summarised in the following.

The first measurement in this research field has been performed in the group of Marti [48]. By acquiring force–distance curves at different temperatures on three polystyrene (PS) samples with different molecular weight, the authors observed a strong increase of the tip-sample adhesion force (jump-off-contact) above a certain temperature, as shown in Fig. 3.26, left panel.

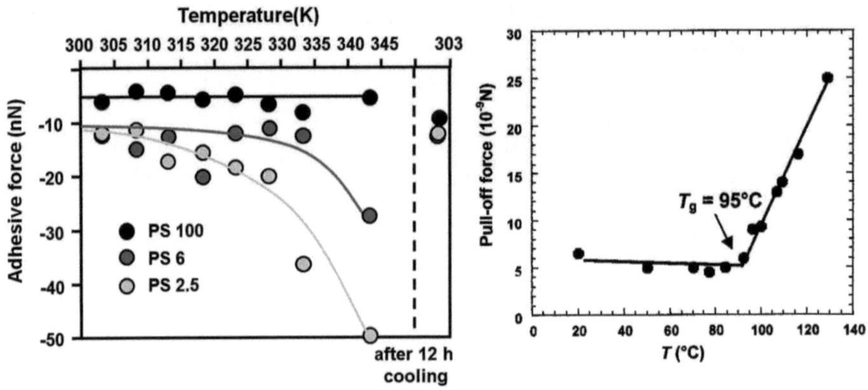

Fig. 3.26 (*Left*) Adhesion force between the AFM tip and three PS samples with molecular weight $M_w = 2.5$ (PS2.5), 6 (PS6) and 100 kg/mol (PS100) as a function of temperature, measured in Ref [48]. For PS2.5 and PS6, the adhesion increases strongly at 320 K and 335 K, respectively, but stays constant for PS100. Adapted with permission from [48]. Copyright 1999. Elsevier (*Right*) Jump-off-contact force versus temperature for one of the PS samples ($M_n = 347$ kg/mol) studied in [49]. Reprinted with permission from [49]. Copyright 2002. American Chemical Society

Bliznyuk et al. [49] have measured the adhesion force of eight PS samples with different mean weight M_n at different temperatures. Additionally, the temperature dependence of two other quantities has been characterised:

1. The slope of the final part of the approach contact line as a measure of sample stiffness.
2. The difference in the piezo displacement between approach and retraction at an arbitrarily chosen force (0.1 μN) as a measure of the hysteresis of the loading–unloading cycle.

As shown in Fig. 3.26, right panel, for the adhesion force, these three quantities change suddenly at T_g. Yet, except for the adhesion, the measured quantities have no physical meaning. Hence, the measurements reported provide a method to estimate T_g via AFM force–distance curves, but yield no information about the temperature or frequency dependence of mechanical properties.

The third work of this small group [46] is reported in detail.

Sample Preparation and Instrumentation Poly(*tert*-butyl acrylate) (PtBuA) was purchased from Scientific Polymer Products (Ontario, NY). The molecular weight of the polymer, determined through a Waters gel permeation chromatograph system, was $M_w = 148$ kDa ($M_w/M_n = 17$). The glass transition temperature of the bulk polymer, measured with a Setaram DSC 92 system, was $T_g = 50$ °C.

Thin films of PtBuA, with thickness smaller than 150 nm, were prepared by spin coating 3 % solution of the polymer in toluene at 500 rpm. The root-mean-square surface roughness was approximately 0.9 nm over 0.5×0.5 μm^2 AFM images of the sample surface. Prior to the AFM measurements study, samples were annealed in a vacuum oven at 120 °C overnight.

The viscoelastic properties of the bulk sample were characterised by measuring the dynamic torsion modulus, $G_{torsion}$, in the temperature range from 60 to 90 °C and the dynamic shear modulus, G_{shear}, in the range 25–60 °C. In both measurements, the oscillatory frequency of the applied torque, ν, has been varied from 0.016 to 16 Hz. The Williams–Landel–Ferry shift factors for the bulk polymer, $\log a_T$, were calculated from these results.

Force–distance curves and adhesion data were obtained with a home-built AFM.

The temperature was controlled manually to ±1 °C with a Peltier heater.

The microscope was equipped with silicon nitride cantilevers (Parks Scientific Instruments, Sunnyvale, CA) with spring constant $k_c = 0.5$ N/m and tip radius $R = 50$ nm.

Measurements were performed inside a glovebox with controlled humidity (<10 %).

The adhesion F_{adh} was measured as the jump-off-contact or pull-off force. When the curves presented a slide-off-contact rather than a discontinuity, F_{adh} was put equal to the maximum pull-off force of the force–distance curve. The values presented in the article are average values from 10 to 30 measurements.

Glass-to-Rubber Transition of PtBuA Figure 3.27A shows the adhesion force F_{adh} measured at different temperatures in the range 30–65 °C at a fixed frequency $\nu = 125$ Hz. F_{adh} increases from 5.7 nN at 30 °C to 26 nN at 65 °C. Artefacts due to the temperature dependence of instrumental parameters, such as the cantilever spring constant k_c, could be excluded through measurements on a silicon sample.

Figure 3.27B shows five isotherms of F_{adh} as functions of the frequency ν. Taking advantage of the time–temperature superposition principle, the five isotherms have been shifted of factors $\log(a_T)$ until they overlap the reference isotherm at $T_{ref} = 55$ °C.

The result is shown in the main panel of Fig. 3.28. The inset shows the shift factors $\log(a_T)$ used in building the master curve and the shift factors determined through the bulk measurements of $G_{torsion}$ and G_{shear}. Since the two datasets are in good agreement, the activation energy governing the glass-to-rubber transition in the bulk is the same as at the surface.

The authors interpret the outcomes of the measurement through the creep function, drawing the conclusion that the increase of adhesion is due to changes of the elastic modulus of the sample with temperature and frequency, which in turn are due to the viscoelastic behaviour of the sample. In fact, the adhesion force, proportional to the work of adhesion W, increases with increasing temperature and decreasing frequency because a viscoelastic sample can be described through an effective surface energy, $W_{eff} = WE_0/E(t)$.

Cappella and Stark [47] have analysed in detail the temperature and frequency dependence of the adhesion force employing the measurements on poly(*n*-butyl methacrylate) (PnBMA) (see Sects. 3.10 and 3.14) and on polystyrene (PS) (see Sects. 3.11 and 3.15).

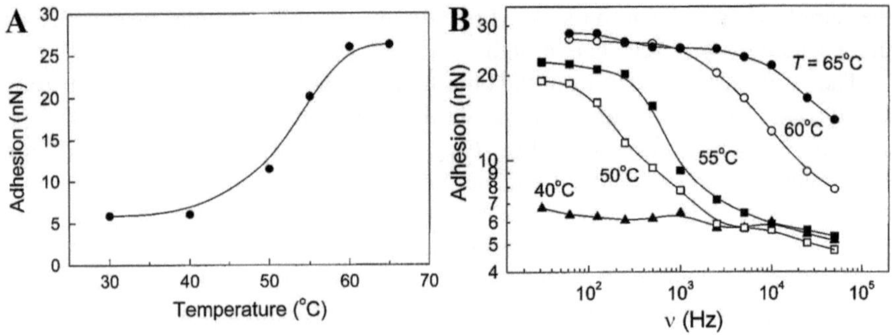

Fig. 3.27 (**A**) Adhesion force measured with a frequency $\nu = 125$ Hz as a function of temperature (**B**) Adhesion force as a function of frequency ν measured at different temperatures, as indicated in the graphic. *Solid lines* in both pictures are only a guide to the eye. Both pictures reprinted with permission from [46]. Copyright 2000. American Chemical Society

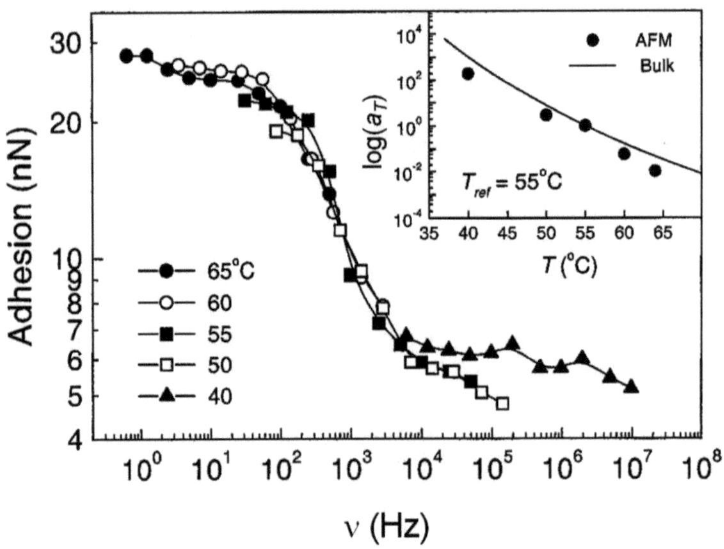

Fig. 3.28 (*Main panel*) Master curve of the adhesion force F_{adh} as a function of the frequency ν. *Solid lines* are only guides to the eye. (*Inset*) Shift factors $\log(a_T)$ as a function of the temperature T. Also the shift factors deduced from the bulk measurements of $G_{torsion}$ and G_{shear} are shown for comparison (*solid line*). Reprinted with permission from [46]. Copyright 2000. American Chemical Society

When measuring the adhesion, it is important to consider the whole force–distance curve, i.e. approach and retraction. Figure 3.29A shows three force–distance curves acquired on PnBMA at different temperatures (30, 43.5 and 51 °C) and at the frequency $\nu = 0.03$ Hz.

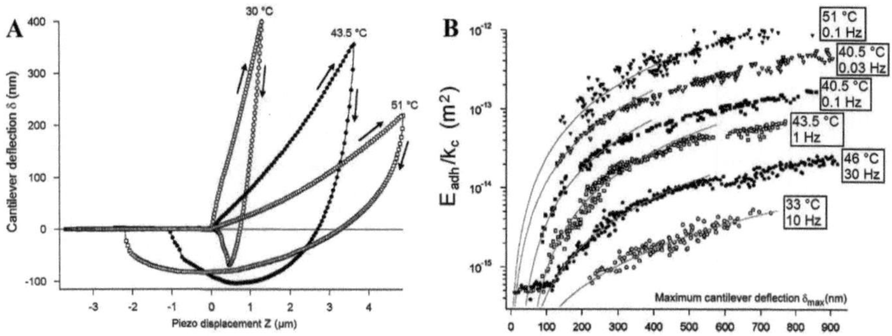

Fig. 3.29 (**A**) Three deflection–displacement curves acquired at $\nu = 0.03$ Hz and three different temperatures, 30 °C (*unfilled circle*), 43.5 °C (*filled circle*) and 51 °C (*unfilled square*). The sample stiffness decreases with increasing temperature, whereas the work of adhesion and the deformation at zero force increase with increasing temperature. (**B**) Work of adhesion E_{adh}/k_c at different temperatures and frequencies, indicated in the tags, versus the maximum cantilever deflection δ_{max}. For $\delta_{max} < \delta_m$ (see Sect. 1.8.3), the work of adhesion is proportional to δ^2_{max} (*grey curves*). Both figures adapted with permission from [47]. Copyright 2005. Elsevier

As already discussed in Sect. 3.10, the stiffness of the sample decreases with temperature. Two important parameters characterising the retraction curves also depend on temperature: the work of adhesion E_{adh}, defined as the area between the axis $F = 0$ and the retraction force–distance curve, and the deformation at zero load D_0, i.e. the intercept between the withdrawal contact line and the axis $F = 0$.

These two parameters increase with increasing temperature. Note that the dependence of the adhesion on temperature would be completely different if the adhesion were measured as the maximum adhesion force; in this case the adhesion would at first increase from $T = 30$ to 43.5 °C and then decrease. This is due to the fact that on such a compliant sample the curves do not present a sharp discontinuity, but rather a slide-off-contact.

It is important to remember that the shape of the approach curves and in particular the stiffness do not change with the maximum force F_{max}. As shown in Sect. 3.10, approach curves acquired at the same temperature and frequency, but with different F_{max}, do overlap with each other. This is not the case for parameters measured from the withdrawal curves, such as E_{adh} and D_0, since the shape of the retraction curves depends indeed on F_{max}. The dependence of the work of adhesion E_{adh}/k_c on the maximum deflection δ_{max} is shown in Fig. 3.29B.

As a consequence of the dependence of E_{adh} on F_{max}, when building a master curve of the work of adhesion as in Fig. 3.28, it is crucial to choose the same maximum force for all curves and to specify it. Another possibility is to use isotherms of the work of adhesion at different maximum loads in order to build master curves of E_{adh} at different maximum loads. This has been done for PnBMA, using the same values $\log(a_T)$ found for Young's modulus (see Sect. 3.14). Figure 3.30 shows two master curves of E_{adh}/k_c, obtained for $\delta_{max} = 200$ nm and $\delta_{max} = 800$ nm. These two master curves confirm that the time–temperature superposition principle can be indeed applied to the work of adhesion, as

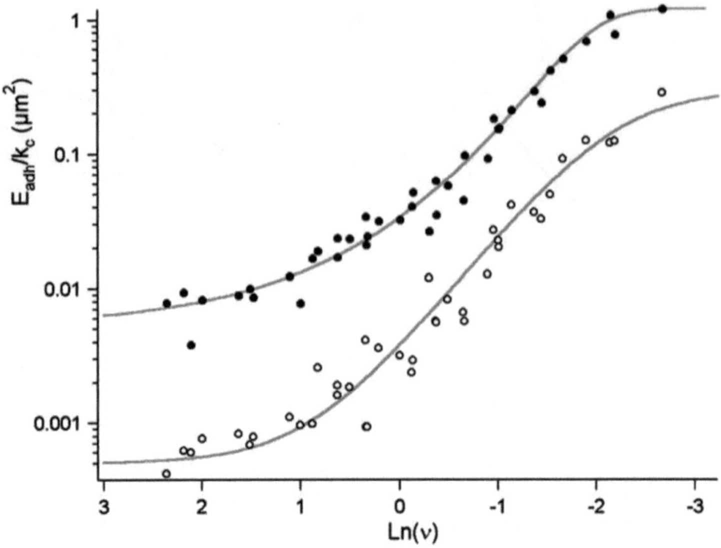

Fig. 3.30 Master curves of the work of adhesion E_{adh}/k_c at $\delta_{max} = 200$ nm (*bottom curve*, *open circles*) and 800 nm (*upper curve*, *points*). The two *grey lines* are only a guide for the eye. The shift values $\log(a_T)$ are the same used for the master curve of Young's modulus shown in Sect. 3.14. Adapted with permission from [47]. Copyright 2005. Elsevier

done by Tsui et al. Moreover, the coefficients of the Williams–Landel–Ferry equation are the same as found for the elastic modulus.

The work of adhesion would not depend on the maximum force if the sample were totally elastic, as can be seen by calculating the work of adhesion in a force–distance curve on the basis of DMT theory.

The work of adhesion can be written as

$$E_{adh} = \left(Z_{joc} - Z_0\right)F_{joc} + \int_{\delta_{joc}}^{0} (Z - Z_0)\mathrm{d}F, \tag{3.6}$$

where Z_{joc} is the piezo elongation at the jump-off-contact, Z_0 is the piezo elongation at $F = 0$ and F_{joc} is the force at the jump-off-contact. Since $Z - Z_0 = \delta + D - D_0$ and the sample deformation at the jump-off contact, D_{joc}, is zero,

$$E_{adh} = \frac{1}{2}k_c\delta_{joc}^2 + \int_{\delta_{joc}}^{0} D\mathrm{d}F. \tag{3.7}$$

With $D = \left(\frac{k_c\delta + 2\pi RW}{E_{tot}\sqrt{R}}\right)^{2/3}$ (see Sect. 1.4.2),

$$\begin{aligned} \int_{\delta_{\mathrm{joc}}}^{0} DdF &= \frac{3}{5}\left(\frac{1}{E_{\mathrm{tot}}\sqrt{R}}\right)^{2/3}\left[(2\pi RW)^{5/3} - \left(k_c\delta_{\mathrm{joc}} + 2\pi RW\right)^{5/3}\right] \\ &= \frac{3E_{\mathrm{tot}}}{5\,R^2}\left(\frac{2\pi R^2 W}{E_{\mathrm{tot}}}\right)^{5/3}, \end{aligned} \tag{3.8}$$

because $k_c\delta_{\mathrm{joc}} = -2\pi RW$. With $a_0 = \left(\frac{2\pi R^2 W}{E_{\mathrm{tot}}}\right)^{2/3}$ the contact radius at zero load and $D_0 = a_0^2/R$,

$$E_{\mathrm{adh}} = \frac{1}{2}k_c\delta_{\mathrm{joc}}^2 + \frac{3}{5}E_{\mathrm{tot}}a_0 D_0^2 = \frac{1}{2}k_c\delta_{\mathrm{joc}}^2 + \frac{6}{5}\pi W a_0^2. \tag{3.9}$$

The work of adhesion can hence be expressed as the sum of two elastic energies resulting from the cantilever (elastic constant k_c and deformation δ_{joc}) and from the sample (elastic constant 6/5 $E_{\mathrm{tot}}a_0$ and deformation D_0).

All parameters in Eq. (3.9), i.e. E_{tot}, k_c, W and R, are independent of the maximum force.

Hence, the increase of the work of adhesion with the maximum deflection shown in Fig. 3.29B is a consequence of the viscoelastic behaviour of the sample and/or of plastic deformations.

The constant value of E_{adh} in the elastic regime (i.e. for $\delta_{\max} < \delta_m$) can be observed in Fig. 3.31, where E_{adh}/k_c versus $\delta_{\max}$ is shown for two PS samples with molecular weight 4 and 62 kDa (see also Sects. 3.11 and 3.15). Since both polymers are in the glassy state at room temperature, the work of adhesion is constant (140 ± 10 nm^2) up to high maximum cantilever deflections also at high temperatures.

When plastic deformations occur (as defined in Sect. 1.9), there is no expression for the work of adhesion as a function of the deformation. Yet two features of the work of adhesion can be assessed:

1. The work of adhesion is proportional to the contact area at zero load $A_0 = \pi a_0^2$ also in presence of plastic deformations. The contact radius at zero load is in turn proportional to D_0. Yet, a_0 and D_0 depend on the maximum force.
2. If $\delta_{\max} \approx \delta_m$, as in the case of the curves on PS shown in Fig. 3.31, plastic deformations are small and the sample behaves almost elastically when the load is released. Hence, the retraction contact line can be approximated with a straight line and D_0 is proportional to $\delta_{\max}$, or, in other words, A_0 is proportional to $\delta_{\max}^2$.

The proportionality between E_{adh}/k_c and $\delta_{\max}^2$ at the beginning of the plastic region can be observed in Fig. 3.29B for PnBMA and in Fig. 3.31 for PS. The transition from the region with constant work of adhesion (i.e. elastic deformations) to the region with $E_{\mathrm{adh}} \propto \delta_{\max}^2$ occurs at lower $\delta_{\max}$ for higher temperatures. For example, such transition is at $\delta_{\max} \approx 80$ nm for $T = 84$ °C and at $\delta_{\max} \approx 200$ nm for

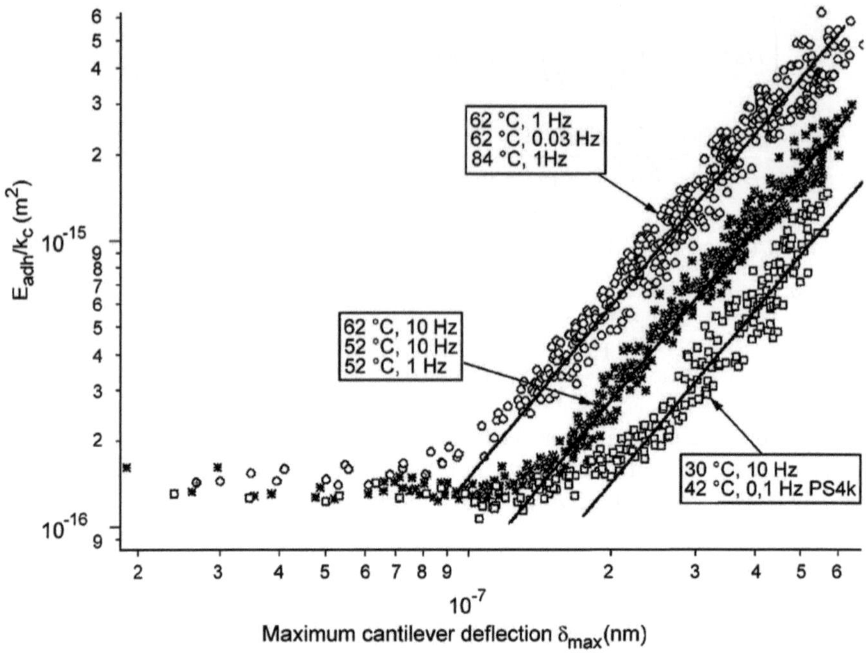

Fig. 3.31 Work of adhesion E_{adh}/k_c versus the maximum cantilever deflection δ_{max} in bilogarithmic scale. Only the data at 42 °C and 0.1 Hz have been acquired on PS with $M_w = 4$ kDa, all other data have been collected on PS with $M_w = 62$ kDa. Temperatures and frequencies of the measurements are listed in the tags. The work of adhesion, constant in the elastic region, is proportional to δ^2_{max} (*straight lines*) at higher forces. Reprinted with permission from [47]. Copyright 2005. Elsevier

$T = 30$ and 42 °C. When measuring E_{adh} through force–distance curves with a fixed maximum deflection, e.g. $\delta_{max} = 150$ nm, and increasing the temperature from 30 to 84 °C, the changes in the work of adhesion would be ascribed to the temperature dependence of the adhesion itself, whereas they are unambiguously due to the temperature dependence of the mechanical properties of the sample and in particular to the temperature dependence of the force $k_c\delta_m$, i.e. the force at which the modulus of the sample goes from E_0 to E_∞ and plastic deformations start occurring, due to the viscoelastic behaviour of the sample.

Considering again the work of adhesion of PnBMA in Fig. 3.29B, it can be seen that, after the region where E_{adh} is constant and the region where $E_{adh} \propto \delta^2_{max}$, at higher forces, there is a third region where $\log(E_{adh})$ is no longer proportional to $\log(\delta_{max})$. This is due to the fact that, when plastic deformations become larger, D_0 is no longer proportional to δ_{max}. Yet, D_0 can be measured for each force–distance curve and E_{adh}/k_c can be divided by D_0^2. The result is shown in Fig. 3.32.

Since all the curves assume the same value once divided by D_0^2, this is a proof that, when plastic deformations take place, the dependence of the work of adhesion

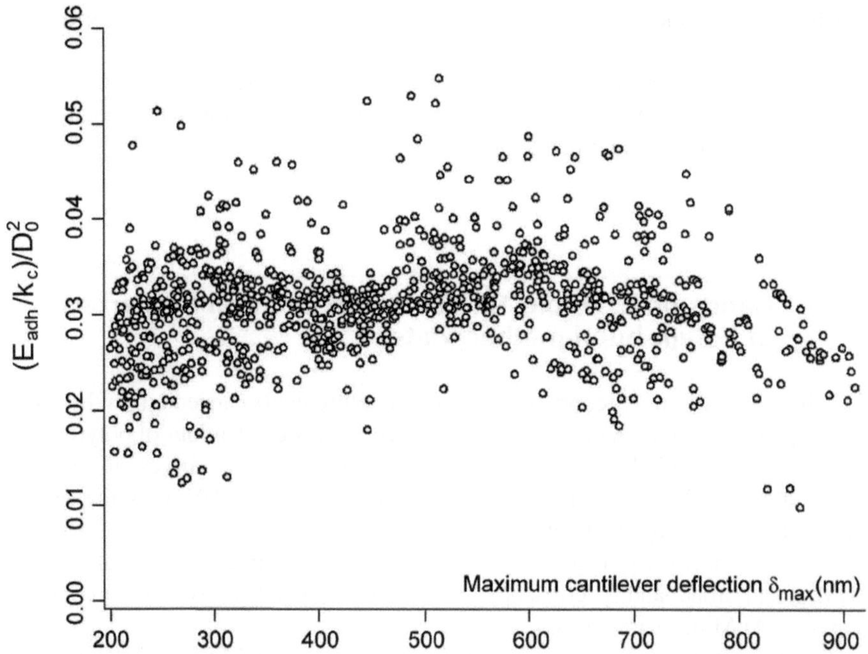

Fig. 3.32 Ratios $(E_{adh}/k_c)/D_0^2$ for the six curves of Fig. 3.29B. Once divided by D_0^2, all the curves assume the same constant value. Adapted with permission from [47]. Copyright 2005. Elsevier

on δ_{max} is due only to the dependence of the contact area at zero load on δ_{max}. In other words, the work of adhesion per unit area, W, does not depend on δ_{max}.

It is important to remember that the curves in Fig. 3.29B and 3.31 have been acquired at different temperatures and frequencies. Once divided by D_0^2, they assume the same value also independently of temperature and frequency. For example, E_{adh}/k_c at $\delta_{max} = 800$ nm varies of two orders of magnitude between 33 and 51 °C, whereas more than 90 % of the values of $(E_{adh}/k_c)/D_0^2$ are between 0.02 and 0.04.

This means that not only the load dependence but also the temperature and frequency dependence of E_{adh} is due to the geometrical term, i.e. to the contact area at zero load, whereas W, i.e. the physical–chemical term, stays constant and depends neither on load nor on temperature and frequency.

Summarising, it can be assessed that the work of adhesion E_{adh} is always proportional to the contact area at zero load A_0. When deformations are elastic, A_0 and E_{adh} are constant. In the presence of plastic deformations as defined in Sect. 1.9, the temperature dependence of E_{adh} is due to the temperature dependence of A_0, which is still proportional to D_0^2, and finally to the temperature dependence of the viscoelastic and plastic properties of the sample. In this case, A_0 and E_{adh} increase with increasing temperature because of the increase of the plastic deformation.

Furthermore, by increasing temperatures, the force at which plastic deformations start occurring becomes smaller.

The important consequence of these results is that the viscoelastic behaviour of a polymer should be studied by characterising the temperature dependence of its mechanical properties and not of its adhesion.

3.14 Hands-on Example 10: Thermomechanical Properties of Poly(*n*-butyl methacrylate)

This hands-on example reports a part of the measurements carried out in the group of Cappella [15]. The characterisation of the viscoelastic behaviour of poly(*n*-butyl methacrylate) (PnBMA) and the fit of deformation–force curves at different temperatures and frequencies has been discussed in Sect. 3.10. Also, the sample preparation is not repeated here.

Yet, it is necessary to give some details about dynamic mechanical analysis (DMA) and broadband spectroscopy measurements, which are shown for comparison with AFM measurements.

DMA measurements were performed with a Netzsch DMA 242 C (Germany), with a temperature range between −60 and 100 °C, a heating rate of 3 K/min and a frequency range between 0.1 and 100 Hz. Complex Young's modulus was calculated from the equation $E^* = (l/A)\ (F^*/\delta^*)$, where δ^* is the amplitude of oscillations (10 μm), A is the surface area of the sample, l its length (9 mm) and the force F is varied, so that the amplitude is constant.

In broadband spectroscopy measurements [50], the complex dielectric function ε^* was measured as a function of the frequency ν in the range $10^{-2} - 10^{7}$ Hz through a high-resolution dielectric spectrometer (Alpha-Analyzer, Novocontrol). The sample temperature was controlled by a nitrogen gas jet cryostat with temperature stability better than 0.1 K (Quadro-System, Novocontrol). Measurements were performed in the temperature range from −20 to 120 °C. To determine the mean relaxation time τ and the relaxation rate at maximal loss ν_m, experimental data were fitted with the function of Havriliak–Negami [51]. Measurements performed show two relaxation processes. The process at higher frequencies and lower temperatures is the β-relaxation corresponding to localised movements of the carbonyl groups; the process at lower frequencies and higher temperatures is the α-process associated with the glass transition.

$D^{3/2}$ curves obtained at different temperatures and frequencies on PnBMA and the fit with the hyperbolic function, Eq. (1.86), have been described in Sect. 3.10.

The fit yields the parameters $\beta + \alpha$ and γ as functions of the temperature T and of the frequency ν. Hence, at each temperature (30, 33, 36.5, 40.5, 43.5, 46 and 51 °C), an isotherm describes the frequency dependence of the fit parameters. Figure 3.33 shows the seven isotherms of the parameter

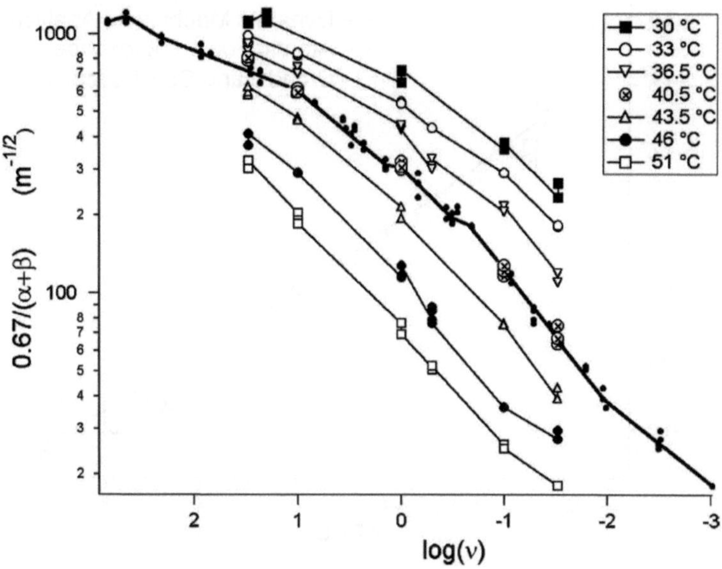

Fig. 3.33 Isotherms of the fit parameter $0.67/(\beta+\alpha)$, proportional to E_∞, as a function of log(ν). Temperatures are indicated in the legend. Also, the master curve obtained by shifting the isotherms until they overlap the reference isotherm at 40.5 °C is shown (•). All *lines* are only a guide for the eye. Reprinted with permission from [15]. Copyright 2005. American Chemical Society

$$\frac{3\left(1-\nu_s^2\right)}{4}\frac{1}{\beta+\alpha}\cong\frac{\sqrt{R}}{k_c}E_\infty(T), \tag{3.10}$$

where ν_s is Poisson's ratio of the sample, k_c the elastic constant of the cantilever, R the tip radius and E_0 and E_∞ are the instantaneous and the infinite elastic modulus.

In Eq. (3.10) it has been assumed that the elastic modulus of the tip is much larger than that of the sample and the term $\left(1-\nu_t^2\right)/E_t$ can be ignored in the expression of the reduced modulus. The factor 0.67 results from Poisson's ratio $\nu_s = 0.33$.

In most of the isotherms in Fig. 3.33, there is more than one point at a certain frequency, resulting from measurements performed at the same temperature and frequency, but on different regions of the sample surface. These measurements confirm that the sample is homogeneous and that the measured mechanical properties are not affected by the topography.

Taking advantage of the time–temperature superposition principle (see Sect. 1.7), the isotherms can be shifted to obtain a master curve of the parameter $0.67/(\beta+\alpha)$. To this aim, chosen a reference isotherm at the temperature T_{ref} (in the present case 40.5 °C), each isotherm is shifted of a shift factor $\log(a_T)$ until it overlaps with the reference isotherm. A shift to the left corresponds for the isotherms in

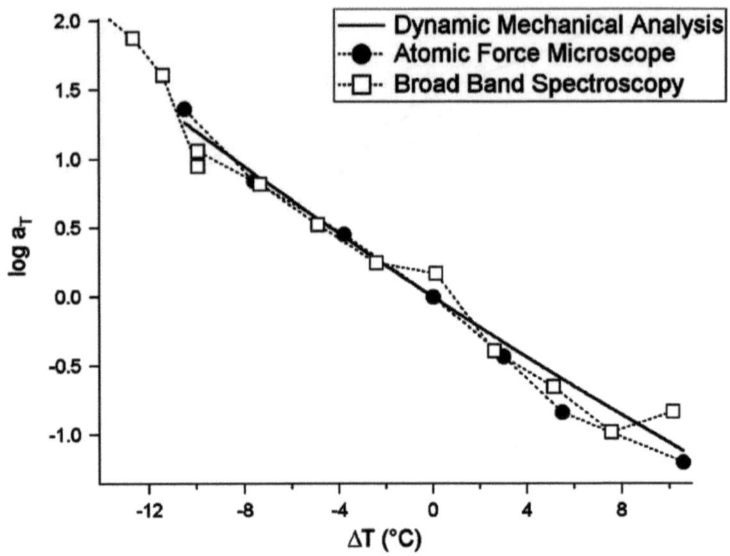

Fig. 3.34 Shift values $\log(a_T)$ versus $\Delta T = T - T_{\text{ref}}$. Three sets of values, obtained through AFM (*full circles*), DMA (*continuous line*), and broadband spectroscopy measurements (*squares*), are shown. Reprinted with permission from [15]. Copyright 2005. American Chemical Society

Fig. 3.33 to an increase of the frequency or a decrease of the probe time and hence to a decrease of the temperature.

The shifted isotherms yield the master curve, showing the dependence of the parameter $0.67/(\beta + \alpha)$ on frequency or, equivalently, on temperature. In Fig. 3.33 also the resulting master curve is shown.

As explained in Sect. 1.7, since all experimental temperatures are between T_g and $T_g + 100$ °C, i.e. 22 and 122 °C, the temperature dependence of the shift values $\log(a_T)$, shown in Fig. 3.34, is described by Williams–Landel–Ferry (WLF) equation.

In Fig. 3.34 also the values of $\log(a_T)$ obtained through DMA and broadband spectroscopy measurements are displayed. AFM results are in very good agreement with those of both techniques.

The calculated values of WLF parameters are $C_1 = 17.3$ and $C_2 = 154$ °C. When $T_{\text{ref}} = T_g$ should be $C_1 = 17.44$ and $C_2 = 51.6$ °C. The large discrepancy in the value of the parameter C_2 is a characteristic of poly(n-alkyl methacrylate)s. The glass transition temperature of these polymers depends on the length of the side groups: longer side groups make the chains more mobile and lead to a decrease of T_g and to a larger transition region between glassy and rubbery plateaus [52]. This peculiarity is due to the unusual strength of the β process, which is stronger than the α relaxation [53, 54].

All parameters obtained from the fit of AFM force–distance curves, in particular γ and $\beta + \alpha$, follow WLF equation with the same values of coefficients C_1 and C_2.

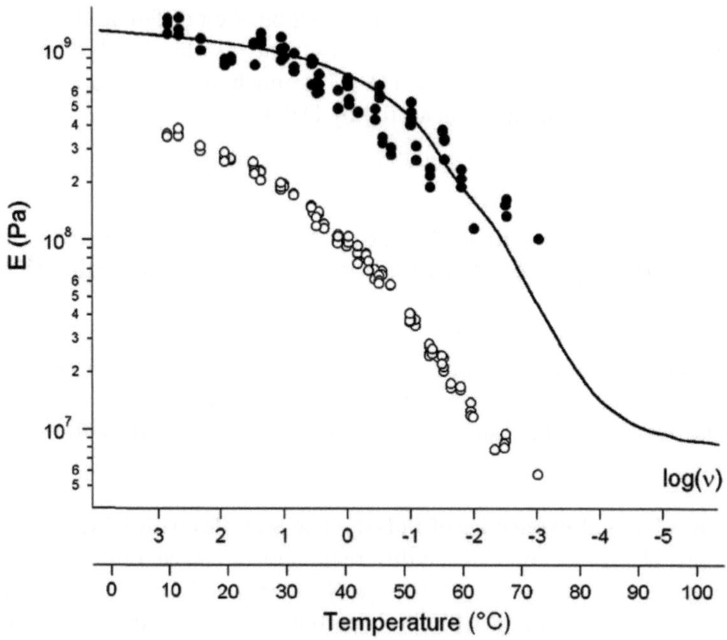

Fig. 3.35 Moduli E_0 (*full circles*) and E_∞ (*open circles*) calculated from AFM force–distance curves together with Young's modulus calculated through DMA (*continuous line*). The moduli are shown versus both the logarithm of the frequency and the temperature, which can be converted into each other by means of William–Landels–Ferry equation. Reprinted with permission from [15]. Copyright 2005. American Chemical Society

Also the adhesion at a given maximum force follows WLF equation with the same coefficients (see previous section and Fig. 3.30).

From the parameters γ and $\beta + \alpha$, it is possible to calculate the instantaneous and the equilibrium moduli of the sample, E_0 and E_∞, since, as already shown,

$$E_0(T) \cong \frac{3\left(1 - v_s^2\right)}{4} \frac{k_c}{\sqrt{R}} \frac{1}{\gamma} \tag{3.11}$$

$$E_\infty(T) \cong \frac{3\left(1 - v_s^2\right)}{4} \frac{k_c}{\sqrt{R}} \frac{1}{\beta + \alpha}. \tag{3.12}$$

Figure 3.35 shows the moduli E_0 and E_∞ calculated from AFM force–distance curves together with Young's modulus measured through DMA. The moduli are shown as a function of both the frequency and the temperature, since frequencies can be converted into temperatures, and vice versa, by means of WLF equation.

Although the modulus E_0 determined through force–distance curves and the modulus $E = \sqrt{E'^2 + E''^2}$ measured by DMA are not the same quantity, the comparison of the experimental results shows that, at least in the glassy temperature interval and at the beginning of the transition region, their difference is smaller than the statistical error in AFM measurements. This is due to the fact that, in the experimental temperature range, (1) E'' is much smaller than E' and (2) E_0 dominates the storage modulus measured by DMA.

3.15 Hands-on Example 11: Thermomechanical Properties of Polystyrene Samples with Different Molecular Weight

This hands-on example reports the measurement of Young's moduli of two polystyrene samples with different molecular weight as a function of temperature. The characterisation of force–distance curves on such samples has been discussed in Sect. 3.11, where also the sample preparation and the instrumentation have been described.

Thermomechanical Properties of Polystyrene Samples with Different Molecular Weight $D^{3/2}$ curves obtained at different temperatures and frequencies on two polystyrene (PS) films with different molecular weight, 4.5 kDa (PS_{4k}) and 62.5 kDa (PS_{62k}), and the fit with the hyperbolic function, Eq. (1.86), have been described in Sect. 3.11.

As a result of the fit, the parameters describing the viscoelastic behaviour of the polymers, i.e. γ and $\beta+\alpha$, are obtained as a function of temperature and frequency:

$$\gamma \cong \frac{3\left(1-v_s^2\right)}{4} \frac{k_c}{\sqrt{R}} \frac{1}{E_0}, \tag{3.13}$$

$$\beta+\alpha \cong \frac{3\left(1-v_s^2\right)}{4} \frac{k_c}{\sqrt{R}} \frac{1}{E_\infty}, \tag{3.14}$$

where v_s is Poisson's ratio of the sample, k_c the elastic constant of the cantilever, R the tip radius and E_0 and E_∞ are the instantaneous and the infinite elastic modulus.

As in the previous hands-on example, in Eqs. (3.13) and (3.14) it has been assumed that the elastic modulus of the tip is much larger than that of the sample and the term $\left(1-v_t^2\right)/E_t$ can be ignored in the expression of the reduced modulus.

Isotherms of both parameters, describing the effect of frequency at a given temperature, are plotted and, taking advantage of time–temperature superposition principle, shifted horizontally in order to get a master curve for E_0 and E_∞. In doing this, values of the shift coefficients $\log a_T$ are obtained. The reference temperature T_{ref} is 54 °C for PS_{4k} and 84 °C for PS_{62k}.

Figure 3.36 shows the master curve obtained for $1/(\beta+\alpha)$, proportional to E_∞, as a function of $\log(v)$ for the PS_{4k} film.

As explained in Sect. 1.7, the temperature dependence of the shift coefficients $\log a_T$ is described by Williams–Landel–Ferry (WLF) equation (Eq. 1.60) in the

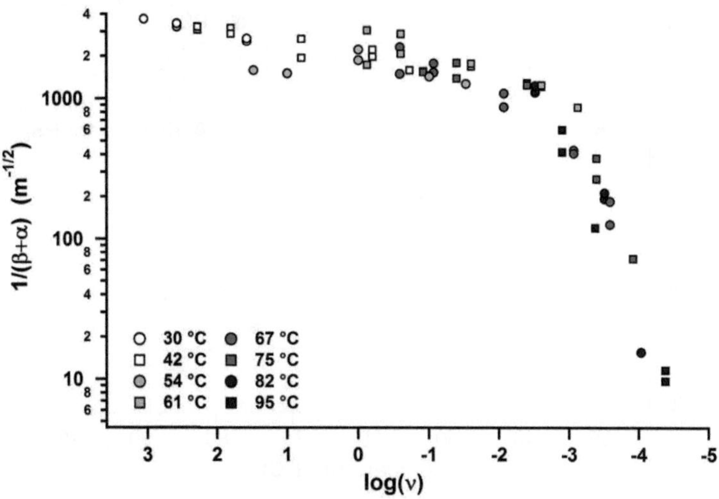

Fig. 3.36 Master curve of the parameter $1/(\beta+\alpha)$, proportional to the equilibrium modulus E_∞, as a function of log(ν). The master curve has been obtained by shifting horizontally the isotherms of $1/(\beta+\alpha)$ of the factors loga_T till they overlap the reference isotherm at 54 °C

range T_g to $T_g + 100$ °C and by Arrhenius equation (Eq. 1.63) for temperatures below T_g. The glass transition temperature of a polymer increases with increasing molecular weight M_n up to a limiting value T_g^∞. Such dependence is described by Fox–Flory relation (Eq. 1.59) [55] in the form $T_g^M = T_g^\infty - K/M_n$.

For polystyrene, T_g^∞ is 100 °C and K is 1.8×10^5 °C. Hence, Fox–Flory relation yields for PS_{62k} $T_g = 97$ °C, in agreement with the value indicated by the provider, and for PS_{4k} $T_g = 55$ °C, which is slightly lower than the value given by the provider.

For PS_{62k}, all experimental temperatures are below $T_g = 97$ °C and only Arrhenius equation has been used to fit the shift coefficients. The glass transition temperature of PS_{4k} is 55–57 °C; hence three experimental temperatures (30, 42 and 54 °C) are below T_g and the other ones (61, 67, 75, 82 and 95 °C) are above T_g. As a consequence, for PS_{4k}, both WLF and Arrhenius equation have been used to fit the shift coefficients. Figure 3.37 shows the shift coefficients versus the temperature for both polymers, fitted with WLF and/or Arrhenius equations.

The parameters calculated from the fits of the shift coefficients are $C_1 = 11$ and $C_2 = 52.8$ K, which are in good agreement with the values expected when $T_{ref} \approx T_g$ [56].

The activation energies are $E_a = 128$ kJ/mol for PS_{4k} and $E_a = 130$ kJ/mol for PS_{62k}.

Dynamic mechanical studies have shown that there are for polystyrene three sub-T_g transitions [57].

The β transition at ca. 325 K with activation energy of about 147 kJ/mol is probably associated with crankshaft-type motions of the PS backbone and vibrational motions of the phenyl rings [58].

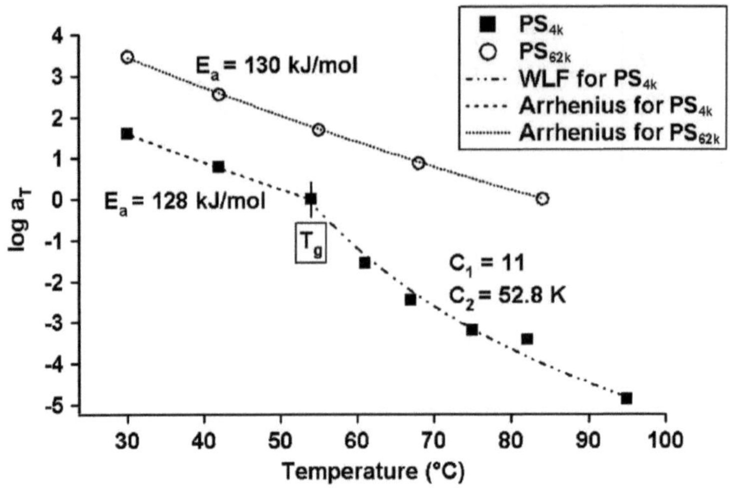

Fig. 3.37 Shift coefficients $\log a_T$, used for the master curve, as a function of temperature for PS_{4k} (*filled squares*) and PS_{62k} (*open circles*). The reference isotherms are those at 54 °C and 84 °C for PS_{4k} and PS_{62k}, respectively. The shift coefficients have been fitted with Arrhenius and WLF equation. The Arrhenius equation has been used for measurements performed at temperatures $T < T_g$ and yields activation energies E_a of 128 and 130 kJ/mol for PS_{4k} and PS_{62k}, respectively. The WLF equation has been used for the measurements performed on PS_{4k} at temperatures $T > T_g$ and yields for the constants C_1 and C_2 the values 11 and 52.8 K. The intersection of Arrhenius and WLF fits permits to estimate the glass transition temperature of PS_{4k} as $T_g = 54$ °C. Reprinted with permission from [41]. Copyright 2005. Elsevier

The γ transition between 130 and 180 K with activation energy of about 42 kJ/mol may be due to the motion of the end groups.

The δ transition at ca. 35 K with activation energy of about 10 kJ/mol is related to partial rotations of the phenyl groups [59].

The activation energy calculated from Arrhenius fit for both samples is in good agreement with the literature value for the β relaxation occurring at 52 °C.

All the parameters obtained from the fit of force–displacement curves, especially γ, follow Williams–Landel–Ferry and Arrhenius equations with the same constants.

In case of PS_{4k}, the intersection of Arrhenius and WLF fit yields an estimation of T_g, 54 °C, which is in good agreement with the known value (57 °C).

Taking advantage of both WLF and Arrhenius equations, frequencies can be converted in temperatures. Figure 3.38 shows the elastic moduli E_0 (left) and E_∞ (right) of PS_{4k} (grey points) and PS_{62k} (white points) in logarithmic scale as a function of temperature.

The results shown in Fig. 3.38 are in agreement with Fox–Flory relation. The elastic modulus E_0 of PS_{4k}, which has the value of about 2 GPa around 10 °C, starts to decrease around $T_g = 57$ °C down to ca. 130 MPa at $T \approx 100$ °C.

In comparison, the modulus of PS_{62K} stays approximately constant in the entire range of experimental temperatures: it is 3.5 GPa from 20 °C up to ca. 70 °C and decreases to just ca. 2 GPa at $T \approx 100$ °C.

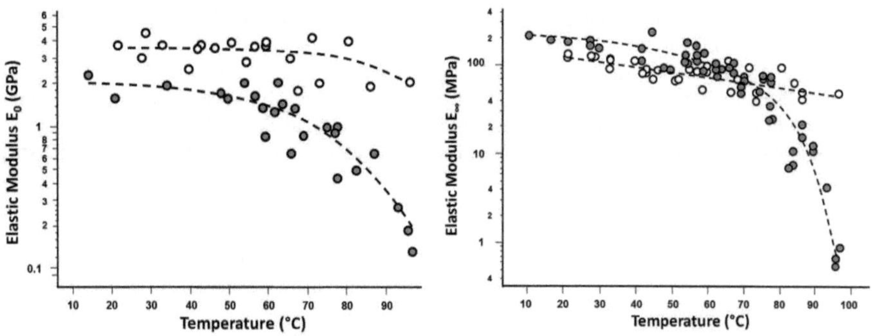

Fig. 3.38 Elastic moduli E_0 (*left*) and E_∞ (*right*) of PS_{4k} (*grey points*) and PS_{62k} (*white points*) as a function of temperature. *Dashed lines* are only a guide for the eye

The infinite elastic modulus E_∞ has a similar dependence on temperature for both samples. For PS_{4k}, it is about 200 MPa around 10 °C and starts decreasing around $T_g = 57$ °C down to ca. 1 MPa at $T \approx 100$ °C, whereas E_∞ of PS_{62k} nearly does not change.

References

1. Carrillo F, Gupta S, Balooch M, Marshall SJ, Marshall GW, Pruitt L, Puttlitz CM (2006) Nanoindentation of polydimethylsiloxane elastomers: effect of crosslinking, work of adhesion, and fluid environment on elastic modulus. J Mater Res 20:2820–2830
2. Kohn JC, Ebenstein DM (2013) Eliminating adhesion errors in nanoindentation of compliant polymers and hydrogels. J Mech Behav Biomed Mater 20:316–326
3. Gupta S, Carrillo F, Li C, Pruitt L, Puttlitz C (2007) Adhesive forces significantly affect elastic modulus determination of soft polymeric materials in nanoindentation. Mater Lett 61:448–451
4. Suriano R, Credi C, Levi M, Turri S (2014) AFM nanoscale indentation in air of polymeric and hybrid materials with highly different stiffness. Appl Surf Sci 311:558–566
5. Griepentrog M, Krämer G, Cappella B (2013) Comparison of nanoindentation and AFM methods for the determination of mechanical properties of polymers. Polym Test 32:455–460
6. Krämer G, Griepentrog M, Bonaccurso E, Cappella B (2014) Study of morphology and mechanical properties of polystyrene–polybutadiene blends with nanometer resolution using AFM and force–distance curves. Eur Pol J 55:123–134
7. Bonaccurso E, Cappella B, Graf K (2006) Local mechanical properties of plasma treated polystyrene surfaces. J Phys Chem B 110:17918–17924
8. Reynaud C, Sommer F, Quet C, El Bounia N, Duc TM (2000) Quantitative determination of Young's modulus on a biphase polymer system using atomic force microscopy. Surf Interface Anal 30:185–189
9. Jee A-Y, Lee M (2010) Comparative analysis on the nanoindentation of polymers using atomic force microscopy. Polym Test 29:95–99
10. Opdahl A, Somorjai GA (2001) Stretched polymer surfaces: atomic force microscopy measurement of the surface deformation and surface elastic properties of stretched polyethylene. J Polym Sci B 39:2263–2274

11. Brostow W (2007) Mechanical properties. In: Mark JE (ed) Physical properties of polymers handbook. Springer, New York
12. Young TJ, Monclus MA, Burnett TL, Broughton WR, Ogin SL, Smith PA (2011) The use of the PeakForce™ quantitative nanomechanical mapping AFM-based method for high-resolution Young's modulus measurement of polymers. Meas Sci Technol 22:125703
13. Silbernagl D, Cappella B (2010) Mechanical properties of thin polymer films on stiff substrates. Scanning 32:282–293
14. Cappella B (2011) Mechanical properties and adhesion of a micro structured polymer blend. Polymers 3:1091–1106
15. Cappella B, Kaliappan SK, Sturm H (2005) Using AFM force-distance curves to study the glass-to-rubber transition of amorphous polymers and their elastic-plastic properties as a function of temperature. Macromolecules 38:1874–1881
16. Chyasnavichyus M, Young SL, Tsukruk VV (2014) Probing of polymer surfaces in the viscoelastic regime. Langmuir 30:10566–10582
17. Du B, Tsui OKC, Zhang Q, He T (2001) Study of elastic modulus and yield strength of polymer thin films using atomic force microscopy. Langmuir 17:3286–3291
18. Cheng X, Putz KW, Wood CD, Brinson LC (2015) Characterization of local elastic modulus in confined polymer films via AFM indentation. Macromol Rapid Commun 36:391–397
19. Lubarsky GV, Davidson MR, Bradley RH (2004) Elastic modulus, oxidation depth and adhesion force of surface modified polystyrene studied by AFM and XPS. Surf Sci 558:135–144
20. Miyake K, Satomi N, Sasaki S (2006) Elastic modulus of polystyrene film from near surface to bulk measured by nanoindentation using atomic force microscopy. Appl Phys Lett 89:031925
21. Chen A, Mu W, Chen Y (2014) Compressive elastic moduli and polishing performance of non-rigid core/shell structured PS/SiO_2 composite abrasives evaluated by AFM. Appl Surf Sci 290:433–439
22. Tsukruk VV, Gorbunov VV, Huang Z, Chizhik SA (2000) Dynamic microprobing of visco-elastic polymer properties. Polym Int 49:441–444
23. Dokukin ME, Sokolov I (2012) On the measurements of rigidity modulus of soft materials in nanoindentation experiments at small depths. Macromolecules 45:4277–4288
24. Hutter JL, Bechhoefer J (1993) Calibration of atomic-force microscope tips. Rev Sci Instrum 64:1868–1878
25. Fischer-Cripps AC (2004) Nanoindentation. Springer, New York
26. Fischer-Cripps AC (2006) Critical review of analysis and interpretation of nanoindentation test data. Surf Coat Technol 200:4153–4165
27. VanLandingham MR (2003) Review of instrumented indentation. J Res Nat Inst Stand Technol 108:249–265
28. Bauman RG, Maron SH (1956) Oxidation of polybutadiene. I. Rate of oxidation. J Polym Sci 22:1–12
29. Shimada J, Kabuki K (1968) The mechanism of oxidative degradation of ABS resin. Part I. The mechanism of thermooxidative degradation. J Appl Polym Sci 12:655–669
30. Raghavan D, Gu X, Nguyen T, VanLandingham M, Karim A (2000) Mapping polymer heterogeneity using atomic force microscopy phase imaging and nanoscale indentation. Macromolecules 33:2573–2583
31. Zhao XM, Xia YN, Schueller OJA, Qin D, Whitesides GM (1998) Fabrication of microstructures using shrinkable polystyrene films. Sens Actuators A 65:209–217
32. Larsson A, Dérand H (2002) Stability of polycarbonate and polystyrene surfaces after hydrophilization with high intensity oxygen RF plasma. J Colloid Interface Sci 246:214–221
33. Cappella B, Wassenberg JR, Heim L-O, Klostermann M, Venzmer J, Bonaccurso E (2014) Mechanical properties of silicone methacrylate microparticles determined by AFM colloidal probe technique. Polymer 55:1209–1216

34. Stolz M, Raiteri R, Daniels AU, VanLandingham MR, Baschong W, Aebi U (2004) Dynamic elastic modulus of porcine articular cartilage determined at two different levels of tissue organization by indentation-type atomic force microscopy. Biophys J 86:3269–3283
35. Vakarelski IU, Toritani A, Nakayama M, Higashitani K (2001) Deformation and adhesion of elastomer microparticles evaluated by AFM. Langmuir 17:4739–4745
36. Schaefer DM, Carpenter M, Gady B, Reifenberger R, DeMejo LP, Rimai DS (1995) Surface roughness and its influence on particle adhesion using atomic force techniques. J Adhes Sci Technol 9:1049–1062
37. Guo D, Li J, Xie G, Wang Y, Luo J (2014) Elastic properties of polystyrene nanospheres evaluated with atomic force microscopy: size effect and error analysis. Langmuir 30:7206–7212
38. Buzio R, Bosca A, Krol S, Marchetto D, Valeri S, Valbusa U (2007) Deformation and adhesion of elastomer poly(dimethylsiloxane) colloidal AFM probes. Langmuir 23:9293–9302
39. Sun Y, Akremitchev B, Walker GC (2004) Using the adhesive interaction between atomic force microscopy tips and polymer surfaces to measure the elastic modulus of compliant samples. Langmuir 20:5837–5845
40. Sun Y, Walker GC (2005) Viscoelastic response of poly(dimethyl siloxane) in the adhesive interaction with AFM tips. Langmuir 21:8694–8702
41. Kaliappan SK, Cappella B (2005) Temperature dependent elastic–plastic behaviour of polystyrene studied using AFM force–distance curves. Polymer 46:11416–11423
42. Attard P (2001) Interaction and deformation of viscoelastic particles. 2. Adhesive particles. Langmuir 17:4322–4328
43. Attard P, Gillies G (2001) Deformation and adhesion of viscoelastic particles: theory and atomic force microscopy. Aust J Chem 54:477–485
44. Gillies G, Prestidge CA, Attard P (2002) An AFM study of the deformation and nanorheology of cross-linked PDMS droplets. Langmuir 18:1674–1679
45. Attard P (2007) Measurement and interpretation of elastic and viscoelastic properties with the atomic force microscope. J Phys Condens Matter 19:473201
46. Tsui OKC, Wang XP, Ho JYL, Ng TK, Xiao X (2000) Studying surface glass-to-rubber transition using atomic force microscopic adhesion measurements. Macromolecules 33:4198–4204
47. Cappella B, Stark W (2006) Adhesion of amorphous polymers as a function of temperature probed with AFM force–distance curves. J Colloid Interface Sci 296:507–514
48. Marti O, Stifter T, Waschipky H, Quintus M, Hild S (1999) Scanning probe microscopy of heterogeneous polymers. Colloids Surf A 1999(154):65–73
49. Bliznyuk VN, Assender HE, Briggs GAD (2002) Surface glass transition temperature of amorphous polymers. A new insight with SFM. Macromolecules 35:6613–6622
50. Kremer F, Schönhals A (2002) Broadband dielectric spectroscopy. Springer, Berlin
51. Havriliak S, Negami S (1966) A complex plane analysis of alpha-dispersions in some polymer systems. J Polym Sci C 14:99–117
52. Hempel E, Beiner M, Renner T, Donth E (1996) Linearity of heat capacity step near the onset of alpha glass transition in poly(n-alkyl methacrylate)s. Acta Polym 47:525–529
53. Kulik AS, Beckham HW, Schmidt-Rohr K, Radloff D, Pawelzik U, Boeffel C, Spiess HW (1994) Coupling of the alpha and beta-processes in poly(etyl methacrylate) investigated by multidimensional NMR. Macromolecules 27:4746–4754
54. Beiner M (2001) Relaxation in poly(alkyl methacrylate)s: crossover region and nanophase separation. Macromol Chem Rapid Commun 22:869–895
55. Fox T, Flory P (1954) The glass temperature and related properties of polystyrene – influence of molecular weight. J Polym Sci 14:315–319
56. Plazek DJ (1965) Temperature dependence of viscoelastic behavior of polystyrene. J Phys Chem 69:3480–3487
57. Fried JR (1996) Sub-Tg transitions. In: Mark JE (ed) Physical properties of polymer handbook. AIP Press, New York

58. Tiller AR (1992) Dielectric-relaxation in polymers by molecular-dynamics simulation. Macromolecules 25:4605–4611
59. Chung CI, Sauer JA (1971) Low-temperature mechanical relaxations in polymers containing aromatic groups. J Polym Sci A2(9):1097–1115

Thin Polymer Films and Polymer Brushes 4

Abstract

The determination of mechanical properties via force–distance curves outlined in the previous chapter, most of all through force–volume measurements, is rather complex and time-consuming. For example, for the determination of the elastic modulus of PnBMA as a function of temperature shown in Sect. 3.14, several thousand curves had to be acquired and analysed. This requires much more time than the measurement and the analysis with, e.g., DMA.

On the other hand, AFM is able to determine the mechanical properties of samples, which cannot be analysed with other techniques. The fundamental difference between AFM and other techniques such as DMA or also nanoindentation is that an AFM enables to determine the local properties of the sample, whereas other techniques give information only about the bulk properties. Hence, an AFM can be employed to study inhomogeneous samples, i.e. samples consisting of two or more phases with one or more interfaces.

There are two wide categories of inhomogeneous samples: thin films and blends. Blends are the object of Chap. 5. In this chapter, thin polymer films and a particular category of them, i.e. polymer brushes, are handled.

4.1 Thin Polymer Films

Theories modelling the deformation and the elastic modulus of thin polymer films on a stiff substrate, i.e. mechanical double layers, have been reviewed in Sect. 1.10.

Thin polymer films are very often prepared through spin coating. The advantage of this technique is that the thickness of the film can be easily tuned and that the resulting polymer layers have a uniform thickness and a flat surface.

In spin coating, the substrate, commonly a glass slide or a silicon wafer, is fixed on a rotating disc. After putting a certain amount of polymer solution onto the centre of the disc, this is rotated, and the solution is distributed uniformly on the substrate. The solvent volatilises during the rotation. Eventually, the substrate can

B. Cappella, *Mechanical Properties of Polymers Measured through AFM Force-Distance Curves*, Springer Laboratory, DOI 10.1007/978-3-319-29459-9_4

be heated in order to obtain a totally solvent-free film. The rotation frequency ω, the solution concentration c_0, and, in case of polymers, the molecular weight M_w determine the thickness of the resulting film [1, 2].

Even being a technique providing uniform, flat, and solvent-free films, spin coating, like other preparation techniques, influences the mechanical properties of the sample. In particular, polymer chains and polymer segments are likely to be stretched and aligned and stretching of chains leads to an increase of the elastic modulus of the polymer [3].

Experiments on thin polymer films are presented in four hands-on examples.

In the first one [4, 5], the feasibility of the theoretical model exposed in Sect. 1.10 is proved by means of a large number of measurements on polymer films with thicknesses in the range 2–190 nm.

The second one [4] illustrates the possibility of determining the thickness of a very thin polymer film (0–30 nm) by means of fitting force–distance curves acquired on it. Such non-destructive technique permits to characterise with high resolution the thickness of a thin polymer layer without scratching it.

In the third hands-on example [6], the position and dimension of glass spheres embedded in a poly(methyl methacrylate) film is determined through force–distance curves. In other words, the analysis of deformations of such model composite material permits to detect the topography of "hidden" elements, which are not measurable through other microscopy techniques.

Finally, the last hands-on example [7] shows measurements on a polymer–polymer mechanical double layer, i.e. polystyrene domains embedded in a polybutadiene matrix. Again, through measurements with a stiff cantilever able to pierce the topmost compliant polymer, the topography of embedded polystyrene domains, not accessible even in Tapping Mode, can be surveyed.

4.2 Hands-on Example 12: Mechanical Properties of Thin Poly (*n*-butyl methacrylate) Films

In this example, measurements of the group of Cappella are reported [4, 5].

Sample Preparation and Instrumentation The poly(*n*-butyl methacrylate) (PnBMA) granulate with molecular weight $M_w = 319$ kg/mol and polydispersity factor $M_w/M_n = 2.58$ was acquired from Scientific Polymer Products (Ontario, NY). Toluene (Uvasol, purity $\geq 99.9\,\%$) was purchased from Merck (Darmstadt, Germany).

As substrates, glass cover slides ($20 \times 20 \times 0.15$ mm^3) from Menzel-Gläser (Braunschweig, Germany) were employed, with roughness of about ± 2 nm.

For the preparation of the double layers, the glass cover slides were cleaned and rinsed with toluene. PnBMA was dissolved in toluene. Polymer films were spin coated from 100 ml of the solutions for 1 min. Solution concentrations and rotation speed are listed in Table 4.1. To make sure to perform the measurements on

Table 4.1 Experimental details of the three measurements

	Measurement I	Measurement II	Measurement III
c_0 (g/ml)	0.08	0.042	0.001
ω (rpm)	2400	1000	2000
R (nm)	25 ± 5	27 ± 5	15 ± 5
k_c (N/m)	37	51	44
t_{bulk} (nm)	430 ± 50	470 ± 50	630 ± 50
E_p (GPa)	3.8	3.14	3.4
t_f (nm)	18, 50, 60, 110, 190	20, 30, 40, 50, 60, 70, 90, 120, 150, 180, 210	2, 3, 4, 5, 6, 8, 10, 12, 15
Δt_f (nm)	± 3	± 5 ($t_f \leq 70$ nm) ± 15 ($t_f \geq 90$ nm)	± 0.5

c_0 is the concentration of the solution for the spin coating, ω is the rotational speed, R is the radius of the AFM tip, k_c is the elastic constant of the cantilever, t_{bulk} is the thickness of the film considered as a homogeneous sample, E_p is the elastic modulus of PnBMA obtained from the fit of the deformation curves on films with thickness t_{bulk}, t_f is the thickness of the measured films, and Δt_f the uncertainty in the measurement of the film thickness

equilibrated and solvent-free films, the samples were dried for ca. one week under ambient conditions.

The most important parameter in the presented measurements is the thickness of the polymer films. The thickness must not only be varied, but also determined with high resolution.

In measurement I, several samples with different uniform film thickness were prepared. To this aim, the solution was diluted by 10 % after the preparation of each sample. The mean film roughness, determined with AFM topography measurements, resulted to be about ± 2 nm. This excludes measurable effects of the sample roughness on the shape of the curves. In order to determine the film thickness t_f, topography images across a scratch were acquired in Tapping Mode at different points of the sample.

In measurements II and III, one film of non-uniform thickness was employed for each measurement. In measurement II, the sample was annealed for one day at 50 °C, in order to promote the dewetting of the substrate. For the determination of the local film thickness t_f, topography images were acquired on surface areas including also the bare substrate. Thanks to the very small mean roughness of the glass slides, the height of these substrate regions can be assumed as the zero thickness and the local thickness is given by the difference between the local height and the height of the substrate regions.

The thickness t_f of the considered films in measurement I and the local thickness in measurements II and III are listed in Table 4.1 together with the uncertainty in their determination Δt_f.

Measurements were performed with a MFP3D microscope (Asylum Research, Santa Barbara, CA). For all force–distance curves in a measurement, the same cantilever (Pointprobe NCL, Nanosensor, Wetzlar-Blankenfeld) was used. The values of the radius R of the silicon nitride tip ($E_t = 245$ GPa and $\nu_t = 0.27$) are listed in Table 4.1 together with the elastic constant of the cantilevers k_c.

The radius R was estimated from deformation curves on glass in agreement with information provided by the manufacturers. The elastic constant was measured using the noise spectrum [8] (see Sect. 2.2.1).

In measurement I force–volume measurements with 10 × 10 force–distance curves at a frequency of 1 Hz on areas of 30 × 30 μm^2 were performed on each sample. In the two other measurements, one single force–volume with 100 × 100 (measurement II) and 50 × 50 curves (measurement III) were performed on an area of 60 × 60 μm^2 (measurement II) and of 2.5 × 2.5 μm^2 (measurement III) with a frequency of 2 Hz (measurement II) and 1 Hz (measurement III).

The second important parameter in the determination of the properties of mechanical double layers is the elastic modulus of the constituents of the samples, in this case glass and PnBMA.

The elastic modulus of glass is known as $E_s = 72$ GPa with Poisson's factor $\nu = 0.27$.

The elastic modulus of PnBMA, E_p, was measured by fitting the curve acquired on the thickest polymer film with Hertz law ($D^{3/2} = \frac{F}{E_{\mathrm{tot}}\sqrt{R}}$). The thickness of the film considered as homogeneous, t_{bulk}, and the values of Young's modulus of PnBMA, E_p, are also listed in Table 4.1.

There is a very good agreement between the results of the three measurements and with results of other measurements performed on spin-coated PnBMA films [9]; nevertheless, these values are higher than the literature value [10, 11], i.e. between 1 and 1.7 GPa.

The higher value of Young's modulus can be explained considering the sample preparation, as spin coating is likely to stretch and align polymer chains and polymer segments. The increase in the elastic modulus through stretching of polymer chains has been shown through measurements on stretched high-density polyethylene [3].

Mechanical Properties of Thin PnBMA Films Figure 4.1 shows in the left column the experimental curves acquired in the three measurements in the form of $D^{3/2}(\delta)$ curves. The film thickness of the polymer films, listed in Table 4.1, is indicated outside the graphs. The curves on substrate (black squares) and polymer films with thickness t_{bulk} (hollow squares), regarded as a homogeneous sample, are also shown and are fitted with Hertz equation $D^{3/2} = \frac{F}{E_{\mathrm{tot}}\sqrt{R}}$ (Eq. 1.19). As expected, the $D^{3/2}$ curves are straight lines, confirming that Hertz theory can be applied and that the tip apex has the shape of a sphere or of a paraboloid.

The experimental curves show that both the deformations and the compliance of the samples increase with increasing film thickness. As already explained in Sect. 1.10, three regimes can be distinguished:

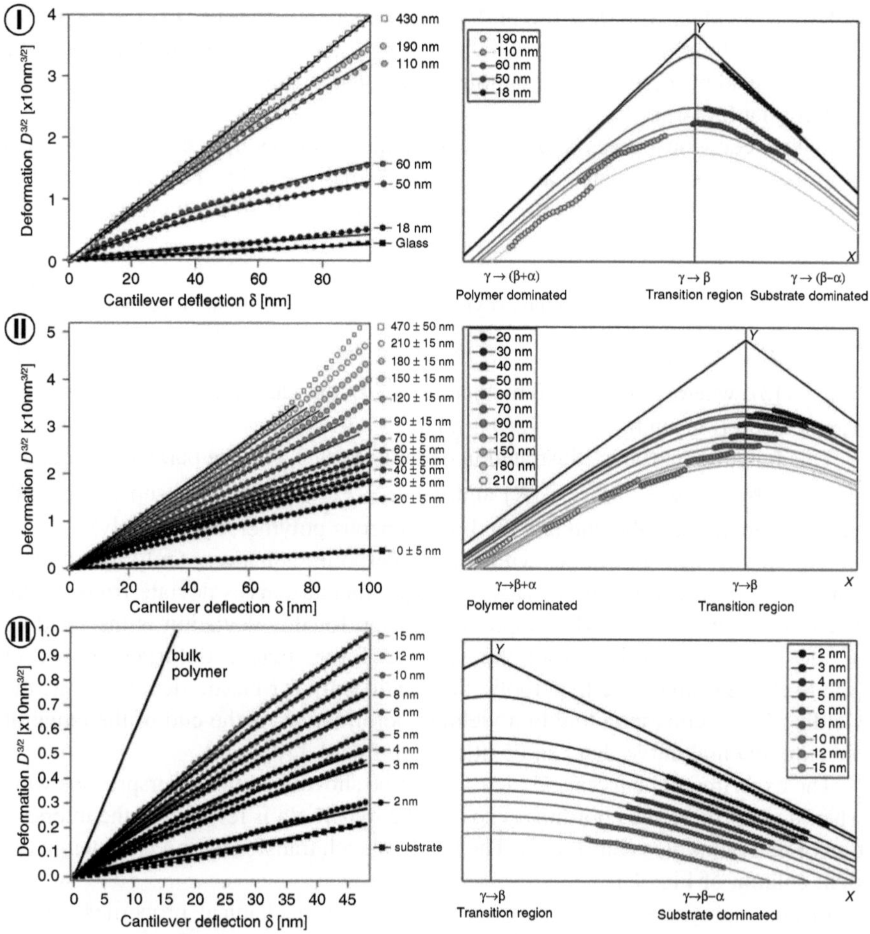

Fig. 4.1 *Left*: Deformation $D^{3/2}$ versus cantilever deflection δ for the three measurements. The thickness of the films, listed in Table 4.1, is repeated in the legends. The curves on substrate and bulk polymer are fitted with Hertz equation. The hyperbolic fits of the experimental curves are shown as *black lines*. *Right*: The same curves shown in the *left panels*, but referred to the interception of the asymptotes. The hyperbolic fit of the curves (*lines*) and the asymptotes $Y = \pm\alpha X$ (*black lines*) are shown too. The experimental curves go through the three regimes, i.e. the substrate-dominated regime in the fourth quadrant, the regime of the mixed properties across the Y axis, and the polymer-dominated regime in the third quadrant, according to the ratio $(\beta - \gamma)/\alpha$. Reprinted with permission from [5]. Copyright 2010. Wiley Periodicals

1. For small values of t_f ($t_f < 40$ nm) and hence large values of D/t_f, the substrate affects the mechanical behaviour of the sample also for small loading forces. As a consequence, the mechanical properties of the sample are dominated by the stiff substrate, and in the whole force range the stiffness of the sample is large and the deformation is small.

2. For large values of t_f ($t_f > 100$ nm) and hence small values of D/t_f, the tip probes only the compliant polymer, which "screens" the substrate and dominates the mechanical properties of the sample. Hence, the stiffness is rather small and the deformation large.
3. In between, the curves show an intermediate behaviour, i.e. the mechanical properties are dominated at small forces (small D/t_f) by the polymer and at large forces (large D/t_f) by the substrate. Hence, the stiffness increases with the loading force.

The $D^{3/2}$ curves acquired on the polymer films are fitted with the hyperbolic function $D^{3/2} = \beta\delta - \varepsilon + \sqrt{\alpha^2\delta^2 + 2\varepsilon(\beta - \gamma)\delta + \varepsilon^2} = \beta\delta - \varepsilon - \alpha\sqrt{(\delta - \delta_m)^2 + \frac{\varepsilon^2}{\alpha^2} - \delta_m^2}$ (Eq. 1.113), where D is the deformation, $\delta = F/k_c$ the cantilever deflection, and the parameter δ_m is given by $\delta_m = -\varepsilon(\beta - \gamma)/\alpha^2$.

This function has been illustrated in detail in Sect. 1.10. The parameters α and β are given by $\beta - \alpha = k_c/\left(E_{\text{tot}}^s\sqrt{R}\right)$ and $\beta + \alpha = k_c/\left(E_{\text{tot}}^p\sqrt{R}\right)$ (the superscripts "s" and "p" denote the substrate and the homogeneous polymer, respectively).

All curves can be fitted with the hyperbolic equation. Only on thicker films ($t_f > 70$ nm), for higher loads, the experimental curves deviate from the fit, as can be seen in the middle graph. The reason for this deviation is the yielding of the polymer. Starting from the yielding force, plastic deformations occur (see Sect. 1.9). Since the hyperbolic fit is valid only for elastic deformations, the yielding force, corresponding to a yielding deformation, is the end of the range of the experimental curve that can be fitted with Eq. (1.113).

The experimental curves and their fits can be shown in a more perspicuous way, when transforming the coordinates so that the hyperbola is referred to the intersection of its asymptotes (Eq. 1.118). The result of such transformation is shown in the right column of Fig. 4.1.

In this representation, the three regimes of the mechanical properties can be easily distinguished. Curves on films with $t_f \geq 110$ nm are in the third quadrant and their properties are dominated by PnBMA; curves on films with a thickness between 30 and 90 nm are situated next to the Y axis and are therefore assigned to the regime of mixed properties; curves on films with $t_f \leq 20$ nm (hence all curves in measurement III) are located in the fourth quadrant and their properties are substrate dominated.

The thickness t_{fit} was calculated from the fitting parameters with Eq. (1.120) as $t_f = c\frac{\varepsilon}{\alpha}\frac{\gamma - (\beta - \alpha)}{(\beta + \alpha) - \gamma}$. The result is shown for all three measurements in Fig. 4.2. For comparison, a straight line with slope 1 going through the origin is shown. The thickness, though varying in a very large range between 2 and 190 nm, can be calculated with good precision.

The errors in the calculated thickness, shown by the error bars, were estimated with the equation $\Delta t_{\text{fit}} = (\partial t_{\text{fit}}/\partial\gamma)\Delta\gamma + (\partial t_{\text{fit}}/\partial\varepsilon)\Delta\varepsilon$. In most cases, the error is so small that the error bars are hidden by the symbols. The error increases with the polymer thickness. This is due to the parameter ε. This parameter is determined

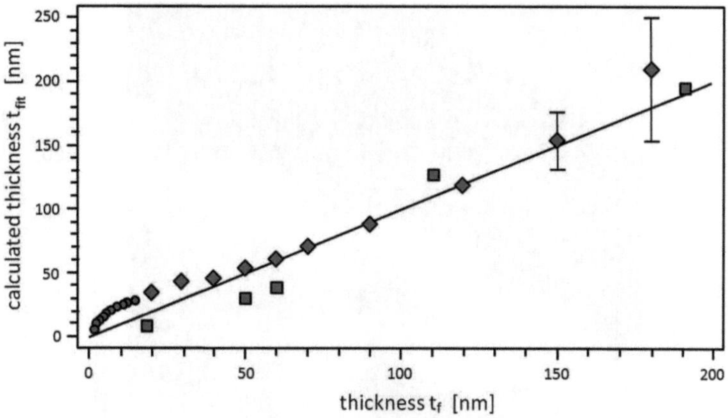

Fig. 4.2 The thickness t_{fit}, calculated from the fit parameters, vs. the measured thickness t_f for measurement I (*squares*), measurement II (*diamonds*), and measurement III (*circles*). The *straight line* is the line $t_{\mathrm{fit}} = t_f$

through the position and width of the transition region. In curves on thick films, the transition region shifts to higher loads outside the measured range. In other words, the small difference between curves on a polymer-dominated sample and on bulk polymer is not sufficient for the precise determination of ε. This effect, together with the maximum load, delimited by the yielding of the polymer, limits the maximum detectable thickness $t_{\max}$.

The proportionality constant $c = 1 \times 10^{-4}$ is the same for all the three measurements and can be considered as specific for the material couple PnBMA/glass. In practice, this means that a force–distance curve, taken on any PnBMA/glass double layer, can be used to determine the film thickness of PnBMA, as the elastic moduli of PnBMA and glass and the constant c are known.

4.3 Hands-on Example 13: Determination of the Thickness of a Dewetted Poly(*n*-butyl methacrylate) Film Through Force–Distance Curves

In this example, the measurements relative to the reconstruction of the thickness of a ultrathin poly(*n*-butyl methacrylate) (PnBMA) film through force–distance curves [4] are summarised. The sample and the instrumentation in the present measurement are the same employed for measurement III in Sect. 4.2.

In order to reconstruct the thickness of an ultrathin PnBMA film, a force–volume measurement with 50 × 50 force–distance curves was performed on a 2.5 × 2.5 μm^2 sample area. The thickness of the PnBMA film ranges from 0, i.e. bare substrate, up to 30 nm. The topography of the surveyed surface area is shown in Fig. 4.3. About

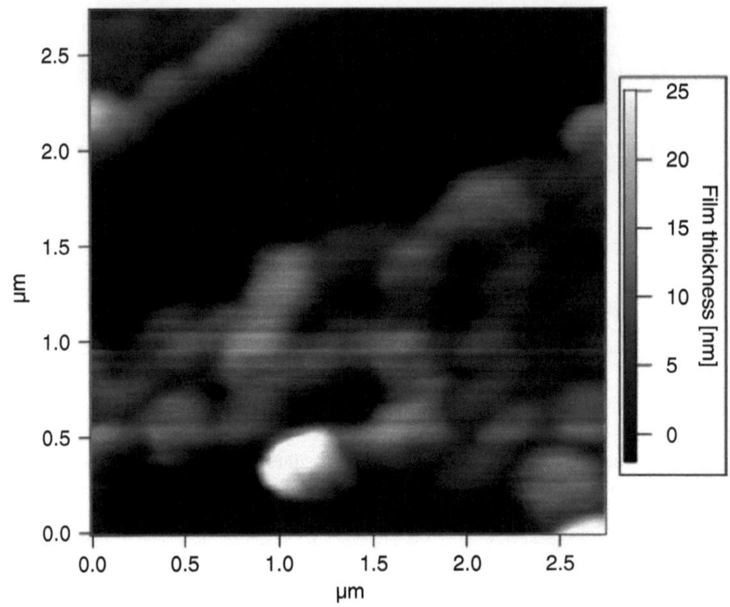

Fig. 4.3 Map of the topography of the examined sample section. The *black* regions represent the uncovered surface of the glass substrate. The *grey* to *white* areas show the regions covered with PnBMA. Reprinted with permission from [4]. Copyright 2007. American Chemical Society

one-third of the surface is uncovered glass substrate. Also, the glass areas are distributed over the whole sample section and the edges to the PnBMA film are easily distinguishable.

The first step of the analysis of the curves is the calculation of the effective stiffness $S_{\text{eff}} = k_s/(k_c + k_s)$ on each point of the area. k_s and k_c are the elastic constant of the sample and of the cantilever (see Sect. 1.2). Since the force–distance curves are spatially resolved, the values of the effective stiffness can be shown in a map, which can be seen in Fig. 4.4A. The agreement between the topography in Fig. 4.3 and the map of the effective stiffness is very good. All glass regions (black regions in Fig. 4.3) can be seen as white regions in Fig. 4.4A. On them, S_{eff} assumes the maximum value, $S_{\text{eff}} = 0.926 \pm 0.012$. Darker regions have a lower effective stiffness and correspond to areas covered with the PnBMA film, whose effective stiffness depends on the film thickness. The two different regions in the map can be distinguished even better in the histogram of S_{eff}, shown in Fig. 4.4B. The values of S_{eff} of polymer and glass are clearly separated. The portion of the histogram acquired on glass (grey bars) can be fitted with a Gaussian curve, and 95 % of the values are in the interval $S_{\text{eff}} = 0.926 \pm 0.012$.

In order to determine the film thickness t_f corresponding to each point of the effective stiffness map, it is necessary to superimpose the topography in Fig. 4.3 and the map of the effective stiffness in Fig. 4.4A. Due to drift during the measurements and to their different resolution, the two images are not congruent

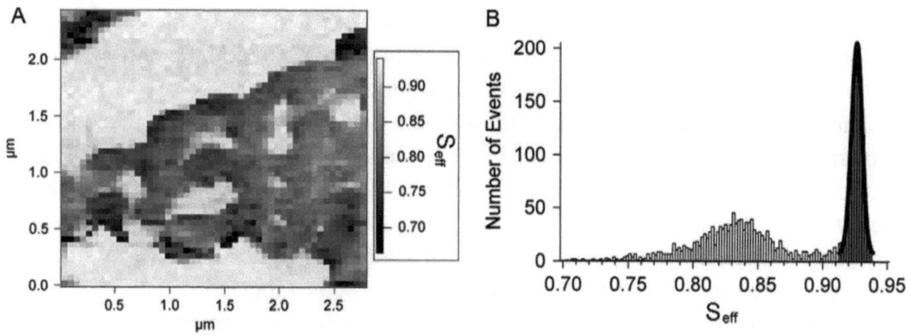

Fig. 4.4 (**A**) Map of the effective stiffness S_{eff} on the examined sample area. The *white* regions correspond to the highest values of S_{eff}, 0.926 ± 0.012, i.e. the effective stiffness of glass. *Darker* spots show a lower S_{eff}, i.e. the effective stiffness of the polymer film, varying with the film thickness. (**B**) Histogram of the effective stiffness S_{eff} with two distinguishable peaks. The peak corresponding to glass surfaces (*grey bars*) is fitted with a Gaussian function, shown as a *black curve*. Reprinted with permission from [4]. Copyright 2007. American Chemical Society

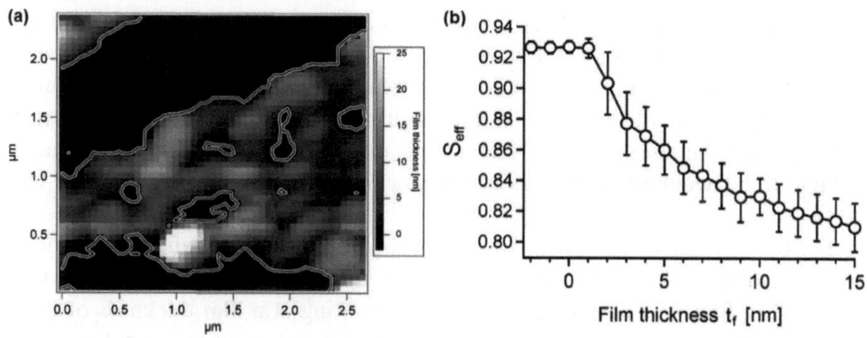

Fig. 4.5 (**a**) Topography image with the same resolution as the map of the effective stiffness. The contour line at $S_{eff} = 0.914$ is superimposed to the topography (*white–black–white line*). This contour line can be regarded as the border between glass and polymer and is a reference for the overlapping of both pictures. (**b**) Effective stiffness S_{eff} of each force–distance curve, grouped in thickness intervals and averaged, plotted versus the film thickness measured in the image in panel (**a**). The error bars show the average deviation of the effective stiffness for each thickness interval. Reprinted with permission from [4]. Copyright 2007. American Chemical Society

and some changes are necessary. To this aim, the glass surfaces can be exploited. The bare glass regions are those with height values between 2 nm and −2 nm and with S_{eff} values between 0.914 and 0.938.

The areas in both images (Figs. 4.3 and 4.4A) with values typical for glass are overlapped. The result of this superposition is shown in Fig. 4.5a.

The topography in this picture is the same as in Fig. 4.3, but its resolution is the same as that of the effective stiffness map. The contour corresponding to $S_{eff} = 0.914$, i.e. the line at which the effective stiffness goes through the value $S_{eff} = 0.914$, corresponding to the border between glass and polymer, is superimposed on the topography. The contour traces not only the largest glass regions, i.e. the approximately diagonal stripe in the top half and the bottom region, but also five smaller spots surrounded by polymer.

Through this superposition the film thickness t_f for each force–distance curve is known, and the effective stiffness can be plotted versus the film thickness. The film thickness was grouped in intervals of 1 nm, and the values of S_{eff} were averaged in each interval. The resulting plot is shown in Fig. 4.5b. The error bars are the average deviation of the effective stiffness over each interval. It can be clearly seen that the stiffness decreases with increasing film thickness t_f, as expected.

The relatively large error bars associated with thicknesses larger than 2 nm show that the correlation of film thickness and S_{eff} is affected by uncertainties, appearing mostly at the edges of the glass areas and due to three phenomena. The first one is the presence of loose polymer residues at the edges of the polymer film, which may engender movements of tip and cantilever, e.g. sticking or slipping, resulting in artefacts in the force–distance curves. Force–distance curves affected by such artefacts, showing a very erratic behaviour, were ignored in the further analysis. The second reason for the anomalous curves at the edges is that a force–distance curve acquired on relatively steep edges yields a false measurement of the stiffness due to changes of the contact area between tip and sample during the indentation (see Sect. 2.3.2). The third reason is that, when a force–distance curve is acquired on the edge, a small deviation in the position of the tip may have important consequences on the measured effective elastic constant, since a curve that is thought to be acquired on glass is actually acquired on the polymer film and vice versa.

For the last step of the reconstruction of the film thickness, deformation curves were calculated. The deformation has been calculated as $D = Z - \delta$ and then averaged in film thickness intervals of 1 nm, starting at a film thickness of 2 nm up to 15 nm, yielding 14 averaged deformation–deflection curves on polymer. All curves taken on a film thickness below 2 nm were averaged to a glass curve. In this way, at least 30 curves were considered for each interval.

Figure 4.6 shows four exemplary deformation–deflection curves, i.e. the curves obtained on glass and on the 2, 3, and 8 nm thick polymer regions. The curves acquired on the polymer film are plotted with their average deviation. Curves whose average deviations do not overlap can be considered as distinguishable. Hence, it can be claimed that AFM indentation is an adequate technique to distinguish a 2 nm thick film from a 3 nm thick film as well as from the substrate.

Deformation curves have been fitted with the hyperbolic function (Eq. 1.113) and the thickness has been calculated as $t_f = c\frac{\varepsilon}{\alpha}\frac{\gamma-(\beta-\alpha)}{(\beta+\alpha)-\gamma}$ (Eq. 1.120).

The result is shown in Fig. 4.7. The circles represent the thickness of each point calculated through Eq. (1.120) versus the thickness measured from the topography (Fig. 4.3); the black line with error bars represents the mean values of the calculated thickness with the average deviation in the 1 nm wide intervals; the empty squares

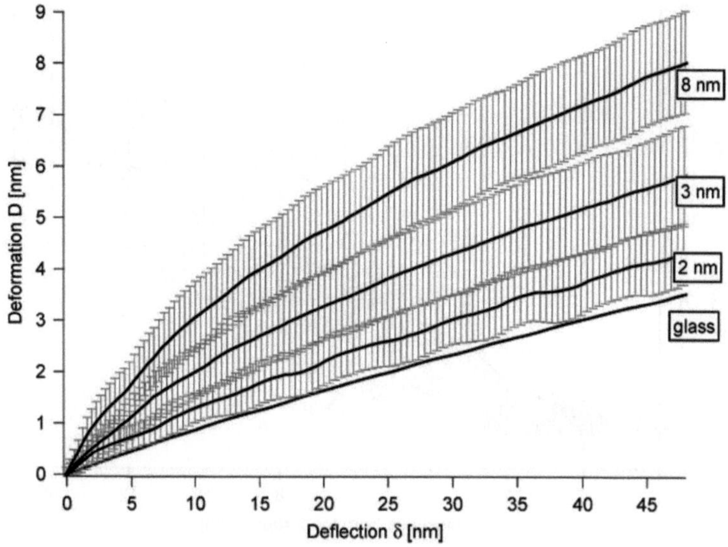

Fig. 4.6 Averaged deformation–deflection curves for the thickness intervals 0, i.e. glass, 2, 3, and 8 nm. The deformation–deflection curves on polymer are plotted with their average deviation to show that the curves are distinguishable. Reprinted with permission from [4]. Copyright 2007. American Chemical Society

are the thickness calculated from the fit of the mean curves, shown in Fig. 4.1 (left column, measurement III). The thickness calculated from the fit parameters is proportional to the thickness obtained from the topography.

The agreement between the mean value of the calculated thicknesses and the thicknesses calculated from the mean curves shows that the two procedures (averaging and fitting) are exchangeable.

The plot of the single points and the error bars of the averages show a rather wide distribution of the values, due to several factors. The most important are, as already explained, the error sources at the edges of the polymer film. Nevertheless, it is possible to calculate the thickness from the fit parameters with an uncertainty of ±3 nm.

Another important reason for the distribution of the calculated thicknesses is that, strictly speaking, the two quantities plotted in Fig. 4.7 are not the same. The thickness measured by means of force–distance curves is actually the thickness of the polymer film, whereas the thickness yielded by the topography is the height difference between a certain point of the topography and the substrate surface. Hence, the two quantities are the same only if the substrate surface is absolutely flat and its height is everywhere zero. This is not the case for common glass substrates. This explains also why the thickness measured from the topography can assume negative values. The mean height of the glass surface has been put equal to zero, and since the mean roughness is ±2 nm, there are also points of the glass surface with "negative" height.

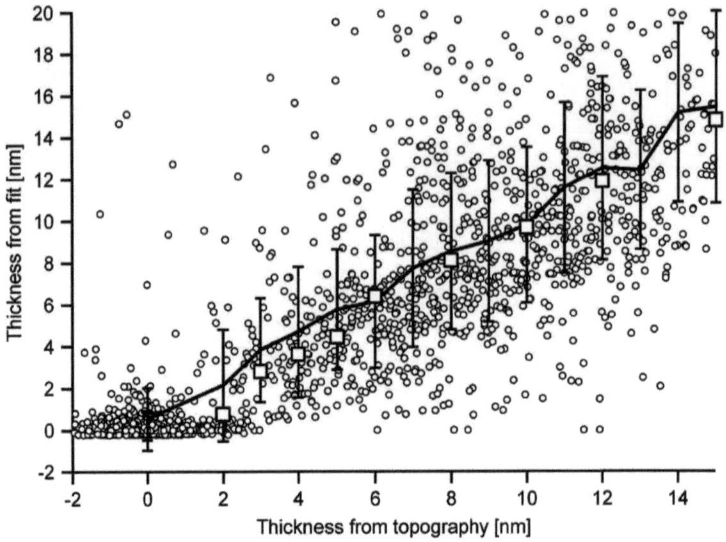

Fig. 4.7 Film thickness calculated from the fit parameters versus the film thickness obtained from the topography. The *circles* show the film thickness obtained from each force–distance curve, with its average curve and average deviation (*black solid line*). Thickness values calculated from the averaged deformation curves are shown by *square markers*. Reprinted with permission from [4]. Copyright 2007. American Chemical Society

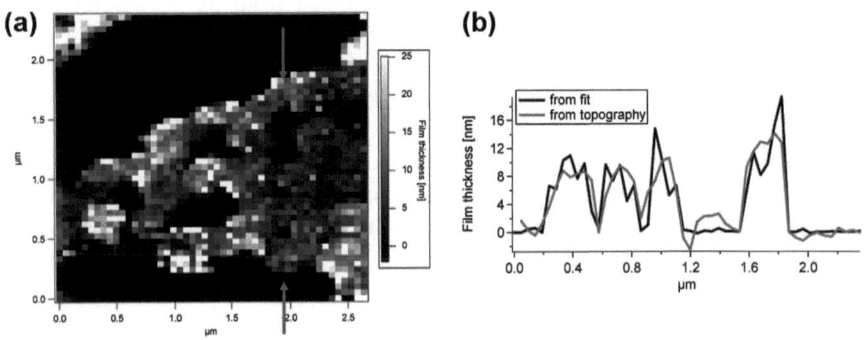

Fig. 4.8 (**a**) Reconstructed topography, in which the film thickness was calculated from the fit parameters of the deformation curves. (**b**) Line profiles of the column marked by the arrows in panel (**a**). The *black line* shows the film thickness calculated from the fit, the *grey line* the corresponding line profile in the topography. Reprinted with permission from [4]. Copyright 2007. American Chemical Society

By fitting the data in Fig. 4.7, the proportionality constant c can be determined and the topography can be reconstructed from the force–distance curves, as shown in Fig. 4.8a.

The reconstructed topography is in very good agreement with the topography in Fig. 4.5a.

Characteristic features of the topography, e.g. the glass spots surrounded by polymer, can be clearly seen in both maps. Figure 4.8b shows the height profile (black line) of one column of Fig. 4.8a (marked by arrows) and the height profile of the corresponding column obtained from the topography (grey line). The five glass intervals can be identified very precisely in the height profile obtained from the fit, and also the height of the polymer intervals is in very good agreement with the topography.

4.4 Hands-on Example 14: Visualisation of Glass Microspheres Embedded in a PMMA Film

In this hands-on example measurements of the group of Cappella [6] are reported.

Sample Preparation and Instrumentation Glass spheres (Spheriglass 3000E CP03, Potters Industries, Eijsden, Netherlands) with a defined distribution of diameters (50 % >35 μm; 13 % <7 μm) were embedded in poly(methyl methacrylate) (PMMA) with $M_w = 120$ kg/mol (PMMA, Plexiglas 7 N, Röhm GmbH, Essen, Germany). PMMA is an appropriate polymer for this sample preparation, since it wets the glass spheres without inner stresses at the glass/polymer interface.

The procedure for the sample preparation is shown schematically in Fig. 4.9. In a first step, glass spheres in toluene (100 μl), after deagglomeration in an ultrasonic bath, were spread on top of a glass slide, the transfer slide, so that, upon evaporation of the solvent, the glass spheres were loosely distributed on the surface of the transfer slide. On a second glass slide, PMMA granulate was molten. In the second step, both glass slides were pressed together and the sample was heated above the glass transition temperature of PMMA ($T_g = 105$ °C). The polymer wetted and embedded the glass spheres, but the glass spheres were still in contact with the transfer slide. In the third step, after cooling down the sample to room temperature,

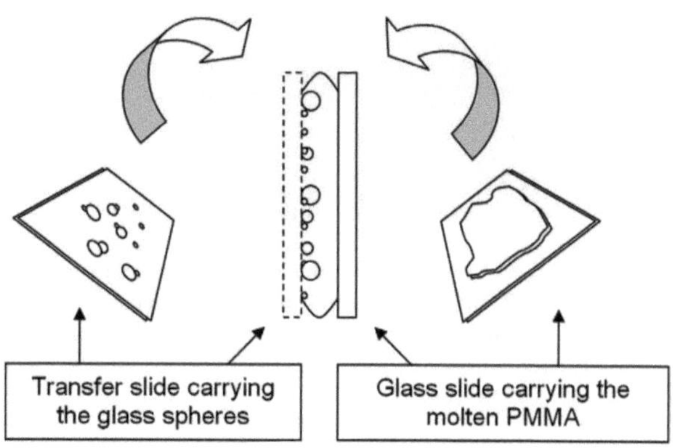

Fig. 4.9 Schematic view of the sample preparation. Reprinted with permission from [6]. Copyright 2009. Elsevier

the transfer slide was removed. As a result, the sample surface to be measured, with the glass spheres underneath the polymer surface, is exposed to air. Since the pressure on the glass slides is removed after cooling down the sample to room temperature, the resulting inner stress cannot be relaxed and the polymer is stiffened.

Measurements were performed with a MFP3D Atomic Force Microscope (Asylum Research, Santa Barbara, CA, USA). The spring constant of the used Pointprobe NCL cantilever (Nanosensor, Wetzlar-Blankenfeld, Germany) was $k_c = 45$ N/m, and the tip radius was about 30 nm.

Measurements of the topography were performed in Tapping Mode and measurements of the mechanical properties in force–volume mode. Two force–volumes with 100 × 100 curves at a frequency of 1 Hz were acquired on two areas of 10 × 10 μm^2, only partially overlapping.

Glass Microspheres Embedded in a PMMA Film In order to localise a sample region with embedded glass spheres, the sample topography was acquired in Tapping Mode. The topography image of a sample region with several spheres embedded is shown in Fig. 4.10.

It can be seen that the molten PMMA has not wetted the glass spheres in the small areas, where they were in contact with the transfer slide, so that small depressions are present on the sample surface. At the border of such depressions, the very thin polymer film was stretched and partially lifted up when the transfer slide was detached, so that the depressions are surrounded by PMMA ribbons or collars.

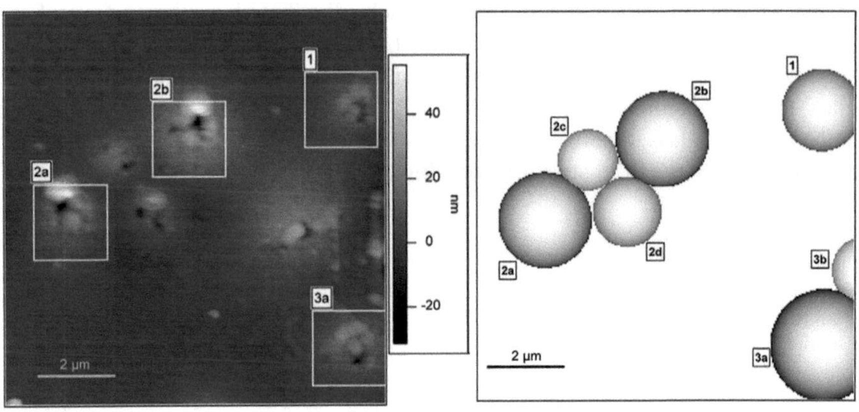

Fig. 4.10 (*Left*) Topography of the examined sample surface with the four regions analysed in detail. (*Right*) Reconstruction of position and dimensions of the glass spheres after removing the polymer matrix in the same area as in the *left image*. Reprinted with permission from [6]. Copyright 2009. Elsevier

Such depressions surrounded by collars indicate the sections of the sample where spheres are close to the surface, i.e., regions where sphere embedded in the polymer can be detected through force–distance curves.

On such regions force–volume measurements have been performed and the curves have been fitted to determine the thickness of the polymer on top of the spheres, thus yielding a "negative" topography of the embedded objects.

In order to check the result of the force–volume measurements, after the acquisition, the polymer was removed by means of plasma etching. After 20 min plasma etching, the glass spheres lay exposed enough to measure their position and size. The reconstructed topography is shown in the right panel of Fig. 4.10. By subtracting the spheres (right panel) from the topography (left panel), the polymer thickness is known. Thus, on each point, the deformation–deflection curve can be related to the polymer thickness and approach curves can be averaged in a certain film thickness interval, as shown in the previous hands-on example. This analysis was performed only on the four sections highlighted in Fig. 4.10, left panel, (1, 2a, 2b, 3a), since the rest of the area included only curves on very thick polymer. Each of the four sections entailed 22 × 22 points of the original force–volumes. The sections 1, 2a, and 2b were measured in both force–volumes and section 3a in only one.

The averaged deformation–deflection curves on polymer layers with different thickness are shown in Fig. 4.11 together with two curves, plotted with light grey markers, acquired on bulk PMMA and on a blank glass sphere. The curves on bulk PMMA and on glass are fitted with Hertz equation $D^{3/2} = \frac{F}{E_{\text{tot}}\sqrt{R}}$ (Eq. 1.19). The fits are shown as black dashed lines. The elastic modulus of the glass substrate, E_s, has been put to 70 GPa in order to determine the sensitivity via Eq. (2.6). With $E_t = 185$ GPa and $\nu_t = 0.28$, the elastic modulus of PMMA is found to be $E_p = 10$ GPa. This value is much higher than the literature value of about 3.5 GPa (see Sect. 3.2). This discrepancy is most likely due to the sample preparation: the pressure applied on both glass slides during the sample preparation engenders stresses at the glass/polymer interface. Such stresses cannot be relaxed since the pressure was removed after cooling down the sample and the glass transition temperature of PMMA is much larger than room temperature.

Curves on very thick polymer coincide with the curve on bulk PMMA. In other words, starting from a certain maximum detectable polymer thickness t_{max}, force–distance curves yield no information about the polymer thickness. t_{max} depends on the tip radius and on the ratio of Young's moduli of polymer and substrate, but also on the yielding force of the polymer (because starting from this force the polymer undergoes plastic deformations, which are not accounted for in the model of deformations of a mechanical double layer) and on the noise of the measurement (because a low noise level permits to distinguish very small deformation differences between a curve on a very thick film and a curve on bulk polymer). In the present case, t_{max} is found to be 375 nm.

All curves can be fitted through the hyperbolic equation (Eq. 1.113) with the fixed parameters α and β, determined through the Hertz fits of glass and bulk

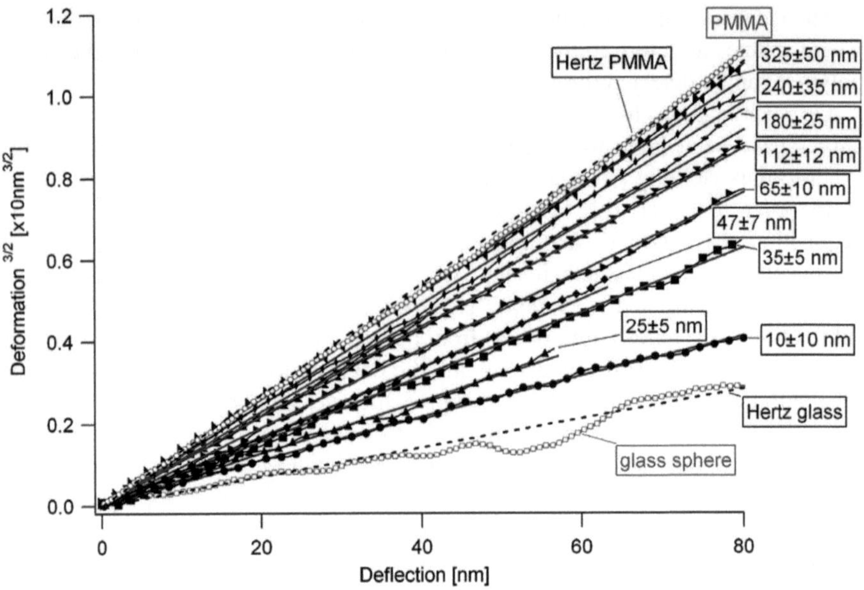

Fig. 4.11 Averaged $D^{3/2}(\delta)$ curves (*markers*) and respective Hertz (*dashed line*) or hyperbolic fits (*solid line*). Reprinted with permission from [6]. Copyright 2009. Elsevier

polymer. The fit parameters are used to calculate the thickness of the polymer film. Equation (1.120), used in the two previous examples, was found to be not adequate for the very large range of polymer thicknesses examined in the present experiment (10–375 nm); in particular, the thickness is not estimated properly when $\gamma \approx \beta + \alpha$, i.e. for very thick polymer films. Hence, Eq. (1.121) has been used in the present work, $t_{\mathrm{fit}} = c\sqrt{\frac{\varepsilon}{\alpha}\frac{\gamma-(\beta-\alpha)}{(\beta+\alpha)-\gamma}}$.

In Fig. 4.12, the polymer thickness calculated through the fit parameters, t_{fit}, is plotted versus the polymer thickness t_f, determined by subtracting the topography of the spheres (Fig. 4.10, right) from the sample topography (Fig. 4.10, left). The plot permits to determine the proportionality factor c as $c = 1.38 \times 10^{-5}\ \mathrm{nm}^{1/2}$.

Using Eq. (1.121) is possible to calculate the polymer thickness, and hence the topography of the embedded spheres, on each point of the four sections in Fig. 4.10 (left panel). The maps of the calculated polymer thickness t_{fit} are shown in Fig. 4.13, superposed for comparison with two contour lines at $t_f = 400$ nm (black line) and $t_f = 50$ nm (grey line). The four spheres, which are embedded in the polymer matrix and cannot be seen in a topography image, can be detected through the described analysis of the force–distance curves.

The grey scale of the maps in Fig. 4.13 ends at $t_{\mathrm{fit}} = 400$ nm. The reason for this is that the maximum detectable polymer thickness is $t_{\mathrm{max}} = 375$ nm. Hence, points with $t_{\mathrm{fit}} > t_{\mathrm{max}}$ are points with undetermined polymer thickness and the polymer thickness on them has been arbitrarily put equal to 400 nm.

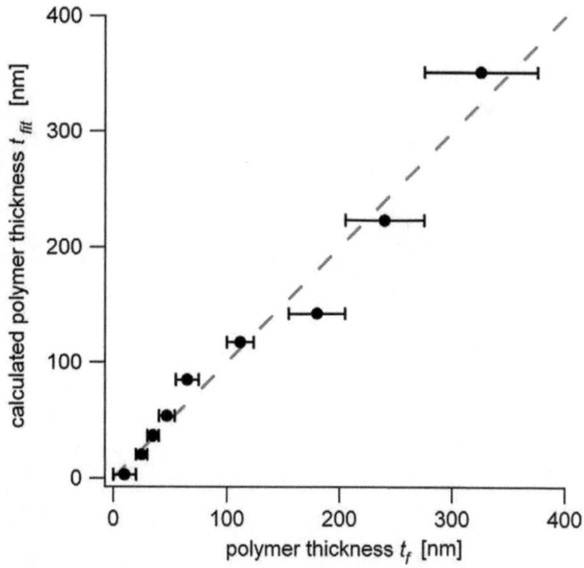

Fig. 4.12 Polymer thickness t_{fit} calculated from fit parameters versus the polymer thickness t_f. *Horizontal error bars* show the width of the polymer thickness intervals, also listed in Fig. 4.11. Reprinted with permission from [6]. Copyright 2009. Elsevier

In the maps, there are also several white points. Such points correspond to curves, which cannot be fitted at all. This is due primarily to two phenomena.

On some points, especially on the collars around the depression on the top of the spheres, the tip gets in contact with loose polymer particles or with polymer layers on top of an air cushion. Typical curves on such points are shown and described in Sect. 2.3.2.

The agreement between the thickness values calculated through the fit and the measured ones is very good. Such an agreement is very noteworthy, because these values are not obtained from averaged curves, as in the previous works, but from single curves, affected by a rather large noise.

4.5 Hands-on Example 15: Force–Distance Curves on a Polymer–Polymer Mechanical Double Layer: Polybutadiene on Polystyrene

This last example of experiments on thin films deals with a part of the measurements reported in [7]. The measurement of the elastic modulus of polybutadiene and its dependence on exposure time to air is outlined in Sect. 3.3; the characterisation of the mechanical properties of the blend is exposed in Sect. 5.6.

Sample Preparation and Instrumentation Polystyrene (PS) with a molecular weight $M_w = 280$ kDa and polybutadiene (PB) with $M_w = 420$ kDa were acquired

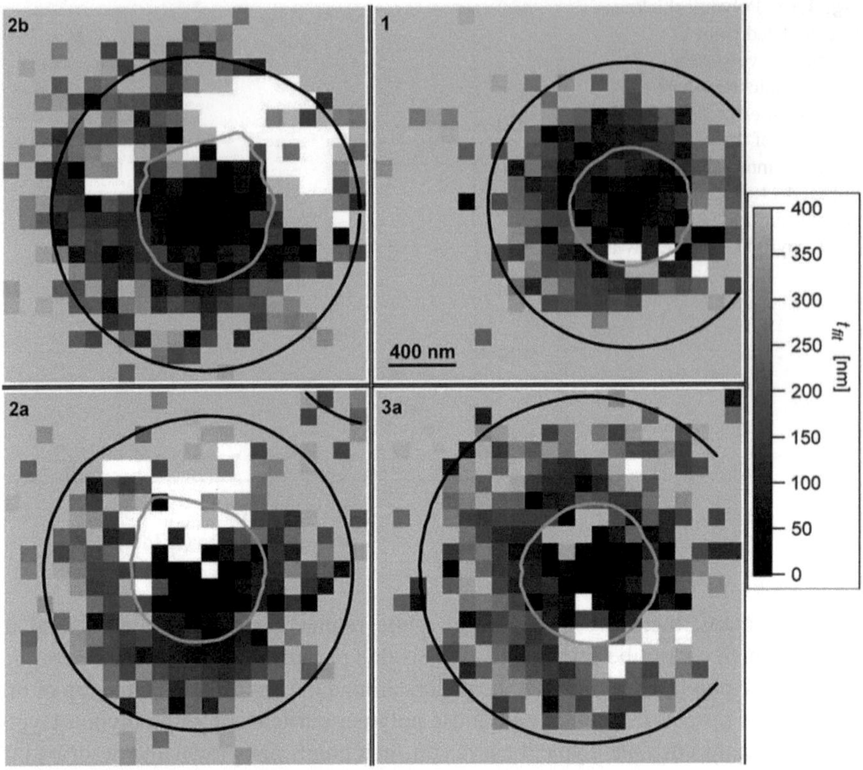

Fig. 4.13 Maps of polymer thickness t_{fit} calculated from the fit parameters on the sample regions 1, 2a, 2b, and 3a, with glass spheres underneath the polymer matrix. Also the contour lines of the polymer thickness at $t_f = 400$ nm (*black line*) and $t_f = 50$ nm (*light grey line*) are shown. Reprinted with permission from [6]. Copyright 2009. Elsevier

from Sigma–Aldrich (St. Louis, MO). The PB contained 36 % cis, 55 % trans, and 9 % 1,2 addition (all percentages are mass fractions).

A PS/PB blend in a 50/50 % weight ratio was prepared by mixing equal quantities of 0.2 % polymer–toluene solutions. Afterwards, the mixed solution was used for spin coating on a glass substrate at 2000 rpm for 30 s to obtain a thin film. The glass substrate was cleaned first with distilled water and acetone in an ultrasonic bath and dried with nitrogen. To avoid oxidation of PB, the blend was stored under inert atmosphere.

Atomic force microscopy measurements have been performed using a MFP3D microscope (Asylum Research, Santa Barbara, CA) equipped with two different cantilevers, a compliant one (elastic constant $k_c = 8$ N/m and tip radius $R = 40$ nm) and a stiff one ($k_c = 74$ N/m and $R = 65$ nm). Cantilever spring constants were measured using the thermal noise method [8].

The tip radii were measured by recording a Tapping Mode image on a test grating with sharp tips (TGT1, NT-MDT, Moscow, Russia). The curvature radius of the tips, provided by the manufacturer, was about 5 nm.

Force–Distance Curves on a PB/PS Mechanical Double Layer During spin coating onto the glass substrate, PS and PB, which are immiscible, undergo phase separation (see Sect. 5.4). The result is a micro-structured blend with PS domains in a PB matrix. Yet, PB wets the PS phase. Since PB, at room temperature, is well above its glass transition temperature, PB chains are very mobile; as a result, the PS domains in the blend are coated with a PB layer. Except for the borders of the PS domains, the thickness of the PB film on top of the PS domains is quite uniform, i.e. 18 ± 3 nm. Hence, the PS domains are a polymer/polymer mechanical double layer, with PB on top and PS as "substrate".

Figure 4.14 shows an averaged $D^{3/2}$ curve recorded with the compliant cantilever on a PB-coated PS domain of the blend (open black circles), together with the averaged $D^{3/2}$ curves measured on homogeneous PB (open grey circles) and on homogeneous PS (filled grey circles). The curve on homogeneous PB has been recorded on the same sample, but on a region of the blend without any PS domain underneath PB. The curve on homogeneous PS has been acquired on a very thick film spin coated on glass with the same parameters as for the blend. The curves on homogeneous PB and on the PB-coated PS domain have been shifted to the right so that the jump-to-contact coincides with the origin.

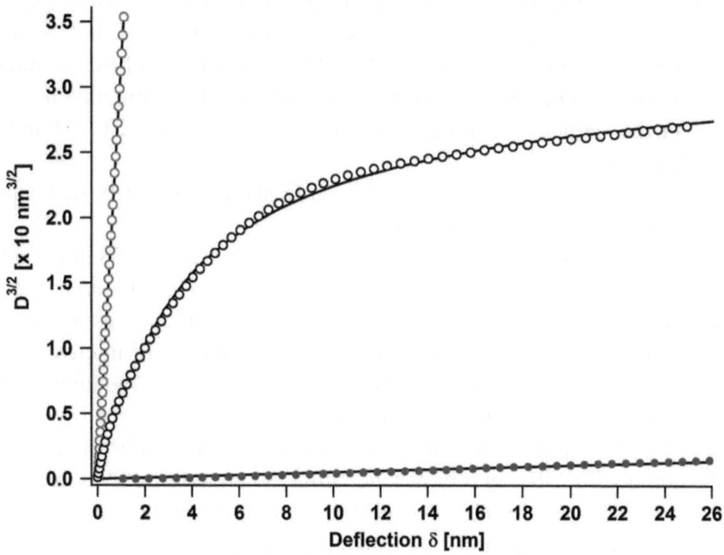

Fig. 4.14 $D^{3/2}$ curves acquired on homogenous PB (*open grey circles*), on homogenous PS (*filled grey circles*), and on the PS/PB mechanical double layer in the blend (*open black circles*). The *black lines* are the fits of the experimental curves with DMT equation (PB and PS) and with the hyperbolic equation (PS/PB)

The $D^{3/2}$ curve on homogeneous PS has been fitted with DMT equation $D^{3/2} = (F + 2\pi RW)/(E_{\text{tot}}\sqrt{R})$ (Eq. 1.24), yielding $E_{PS} = 7 \pm 0.7$ GPa. This value is higher than the literature value (about 3 GPa, compare Sect. 3.4). The stiffening is due to the spin coating used for the sample preparation (see also the preceding hands-on examples in this section). The $D^{3/2}$ curve on homogeneous PB has also been fitted with DMT equation, yielding $E_{PB} = 7.3 \pm 0.3$ MPa. Hence, PB in this sample is already cross-linked due to oxidation and its elastic modulus is higher than that of fresh PB (see Sect. 3.3). Moreover, it should be kept in mind that this value has been obtained with a DMT fit, and not with JKR equation, which should be used for PB (see Sect. 3.3). The use of DMT theory is due to the fact that, in the JKR theory, $D^{3/2}$ is not proportional to the force, whereas in the DMT theory it is proportional to the force shifted of the quantity $2\pi RW$. Hence, the limit of a $D^{3/2}$ curve on a mechanical double layer for very thick films, $(\beta + \alpha)\delta$, is the $D^{3/2}$ curve on homogeneous polymer, provided it is shifted of the quantity $2\pi RW$, as in Fig. 4.14.

The $D^{3/2}$ curve on the mechanical double layer has been fitted with the hyperbolic function in Eq. (1.113) (black line). The fit, obtained with the values of α and β given by the moduli of PB and PS, reproduces quite well the experimental curve.

For $D = 0$, the slope of the blend curve is lower than the slope of the PB curve, i.e. γ is considerably lower than $\beta + \alpha$. This means that the PS substrate affects the elastic modulus of the double-layer system from the very beginning of the indentation. This is a consequence of the rather low thickness of the PB film (18 ± 3 nm). The influence of the PS substrate on the mechanical double layer increases with further indentation. Finally, when the covering PB film is pierced or completely compressed, the elastic modulus of the system is equal to that of the PS substrate.

The thickness of the PB layer on top of the PS domain cannot be determined with these measurements, because PB has almost everywhere the same thickness and the constant c in the semiempirical equation of the thickness (Eqs. 1.120 and 1.121) cannot be determined.

Force–distance curves acquired on the same sample with the stiff cantilever ($k_c = 74$ N/m) show very interesting features.

Figure 4.15 shows three force–distance curves recorded on a PS domain (A), on the PB matrix (B), and inside a cavity in a domain (C). Approach curves are plotted in black and withdrawal curves in grey. The ranges of the curves are not shown completely, as only the region around the jump-to-contact is of interest.

All curves show double-layer behaviour. Immediately after the jump-to-contact the tip penetrates the PB layer, since the stiffness of the cantilever is much higher than that of PB. The slope of the curve is about 0.05 k_c. After a bend, in the second part of the contact line, the tip is either indenting the underlying PS domain (curves A and C) or pushing against the glass substrate (curve B). In the first case, the slope is between 0.5 k_c and 0.65 k_c and the contact lines do not overlap, indicating that the substrate is plastically deformed. In the second case, the slope is nearly equal to k_c and approach and withdrawal contact lines overlap.

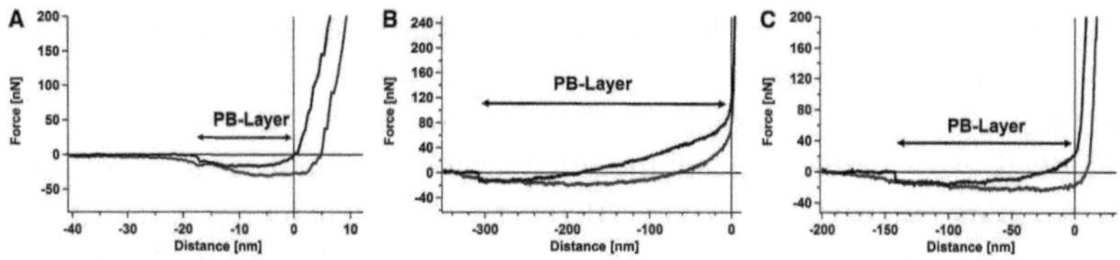

Fig. 4.15 Three force–distance curves recorded on a PS domain (**A**), the PB matrix (**B**), and a hole in the domain (**C**). The PB layer thickness is shown by a *double-headed arrow*. The approach curves are *black* and the withdrawal *grey*. The ranges of the curves are not shown completely. Adapted with permission from [7]. Copyright 2014. Elsevier

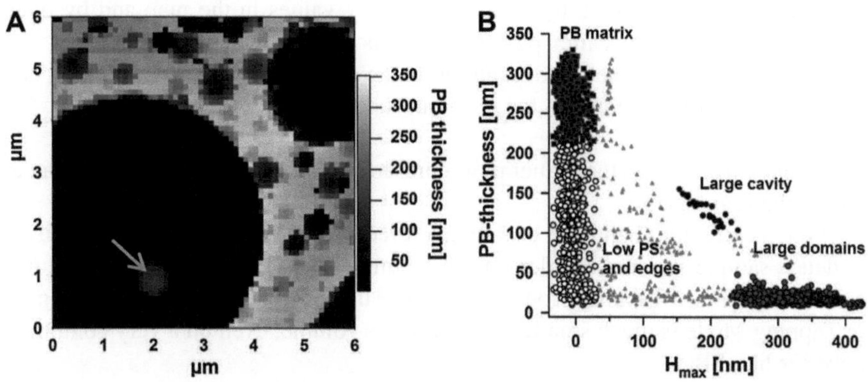

Fig. 4.16 (**A**) Map of the PB thickness in the blend. The layer thickness is small on the PS domains, intermediate in the large hole and over the hidden PS domains, and large in the matrix. (**B**) Scatter plot of the PB thickness versus the height over the glass substrate at which the maximum force is reached, H_{max}. Four different regions can be identified: the PB matrix (*black squares*), the high PS domains (*grey filled circles*), the large cavity (*filled black circles*), and, as a mixed region, the low PS domains and the edges of the higher domains (*open black circles*). The *grey triangles* represent data points which do not belong to any of the four groups. Reprinted with permission from [7]. Copyright 2014. Elsevier

The thickness of the PB layer is indicated in the graphs through double-headed arrows. It can be measured as the Z distance between the jump-to-contact and the bend in the contact line. It is 18 nm for curve A, 300 nm for curve B, and 130 nm for curve C. Moreover, curve C indicates that the cavity does not go through the whole PS domain down to the glass substrate.

By calculating the PB thickness t_{PB} for all 60 × 60 curves of a force–volume, a map of t_{PB} was created (Fig. 4.16A). Panel B of Fig. 4.16 is the corresponding scatter plot of t_{PB} versus H_{max}, i.e. the height over the glass substrate (assumed to be at height zero) at which the maximum force was reached.

Four different regions can be identified in both the map and the scatter plot:

1. The light grey region in the map and the black squares in the scatter plot represent the PB matrix. On it is $t_{PB} = 275 \pm 20$ nm and $H_{max} \approx 0$ nm, since the tip pierces the very compliant PB film down to the glass substrate.
2. The black areas in the map and the filled grey circles in the scatter plot are high PS domains emerging from the PB matrix. On them is $t_{PB} = 18 \pm 3$ nm and $H_{max} = 250$–400 nm. Hence, the tip does not pierce the stiffer PS and does not reach the glass substrate.
3. The large cavity, highlighted in the map by an arrow, corresponds to the filled black circles in the scatter plot. On it is $t_{PB} = 100$–150 nm and $H_{max} = 150$–200 nm. Also in this case the tip does not reach the glass substrate.
4. Low PS domains, not emerging from the PB matrix, and the edges of the larger domains are represented by intermediate grey values in the map and by open black circles in the scatter plot. On them is t_{PB} between 20 and 200 nm and $H_{max} \approx 0$ nm, since the tip, depending on the dimension of the PS domains, can pierce the PS layer down to the glass substrate.

The grey triangles in the scatter plot represent data points which do not belong to any of the four groups.

As in the previous hands-on example, a force–volume measurement can be used to detect sample structures underneath the sample surface. Such sample structures cannot be seen in a Tapping Mode image, since the indentation of the oscillating tip in Tapping Mode is much smaller than in force–volume, even on a very compliant polymer like PB.

4.6 Polymer Brushes

In the next section, two experiments [12, 13] on polymer brushes are reported in one hands-on example.

Since the thickness of a polymer brush cannot be larger than the chain length (see Sect. 1.5), polymer brushes are inherently thin films.

Two different approaches can be employed to fabricate polymer brushes onto a substrate: the "grafting-from" and the "grafting-to" approach [14].

In the grafting-to approach, end-functionalized polymers are chemically attached on a substrate, which is usually also functionalized. Due mainly to steric constraints, commonly only small amounts of polymer (in the range of 2–10 mg/m^2) can be attached onto the substrate [12].

In order to increase the grafting density, the "grafting-from" approach can be used. The grafting-from approach implies the attachment of a radical initiator onto a substrate via SAMs, followed by surface-initiated polymerization to form the brush. With this method, larger grafting densities and brush layer thicknesses are achievable. Yet, the grafting-from approach involves often complicated synthesis in several steps. This is the case for the complex sample preparation in the two works presented in the next section.

The presented experiments show unambiguously the possibility of determining the mechanical properties of polymer brushes and to compare them to those of bulk samples of the same polymer. Furthermore, in the second part of the hands-on example, the characterization of the mechanical properties of a binary polymer brush yields extensive information about the morphology of the sample.

4.7 Hands-on Example 16: Elastic Modulus of Poly(styrene-*co*-pentafluorostyrene) and Poly(methyl acrylate) Polymer Brushes

In the present hands-on example, two experiments of the group of Tsukruk [12, 13] are reported.

Sample Preparation and Instrumentation The rather complicated procedure used to prepare the polymer brushes characterised in both articles is schematically shown in Fig. 4.17 together with the structure formulas of poly(styrene-*co*-pentafluorostyrene) (PSF) and poly(methyl acrylate) (PMA) (upper part, respectively, left and right formula).

The procedure involves four steps:

1. Modification of the substrate ([100] silicon wafers) with (3-Glycidoxypropyl) trimethoxysilane (GPS) [15, 16].
2. Treatment with 1.5 % ethylendiamine in ethanol for 1.5 h.
3. Treatment with a solution of 0.66 g of the acid chloride derivative of 4,4′-azobis (4-cyanopentanoic acid) (ABCPA) and 0.36 mL of triethylamine in 50 ml of CH_2Cl_2 for 2 h.
4. “Grafting-from” polymerization.

For this last step, the Si wafers with the grafted azo initiator were placed under argon atmosphere in a glovebox in a reactor together with the monomers [73 g (0.70 mol) of S and 20 g (0.10 mol) of FS in 100 g of THF for PSF and 50 g (0.58 mol) of MA in 50 g of toluene for PMA]. The reactor was immersed into a water bath (60 °C) for 12 h.

Further details about the reactions route and the sample preparation are given in [12, 13].

In the first article [12], additional spin-coated PMA and PSF layers were prepared from toluene (for PSF) and acetone (for PMA) solutions, in order to compare the mechanical properties of brush layers with those of polymer films in which the chains possess the same conformation as in the bulk state.

In both articles, the film thickness was measured by a Compel automatic ellipsometer (InOm Tech, Inc.) and with AFM topography images over a scratch.

The AFM measurements were performed with the Dimension 3000 and the Multimode atomic force microscopes (Digital Instruments, Santa Barbara, CA). The microscopes were equipped with silicon or silicon nitride tips with radii

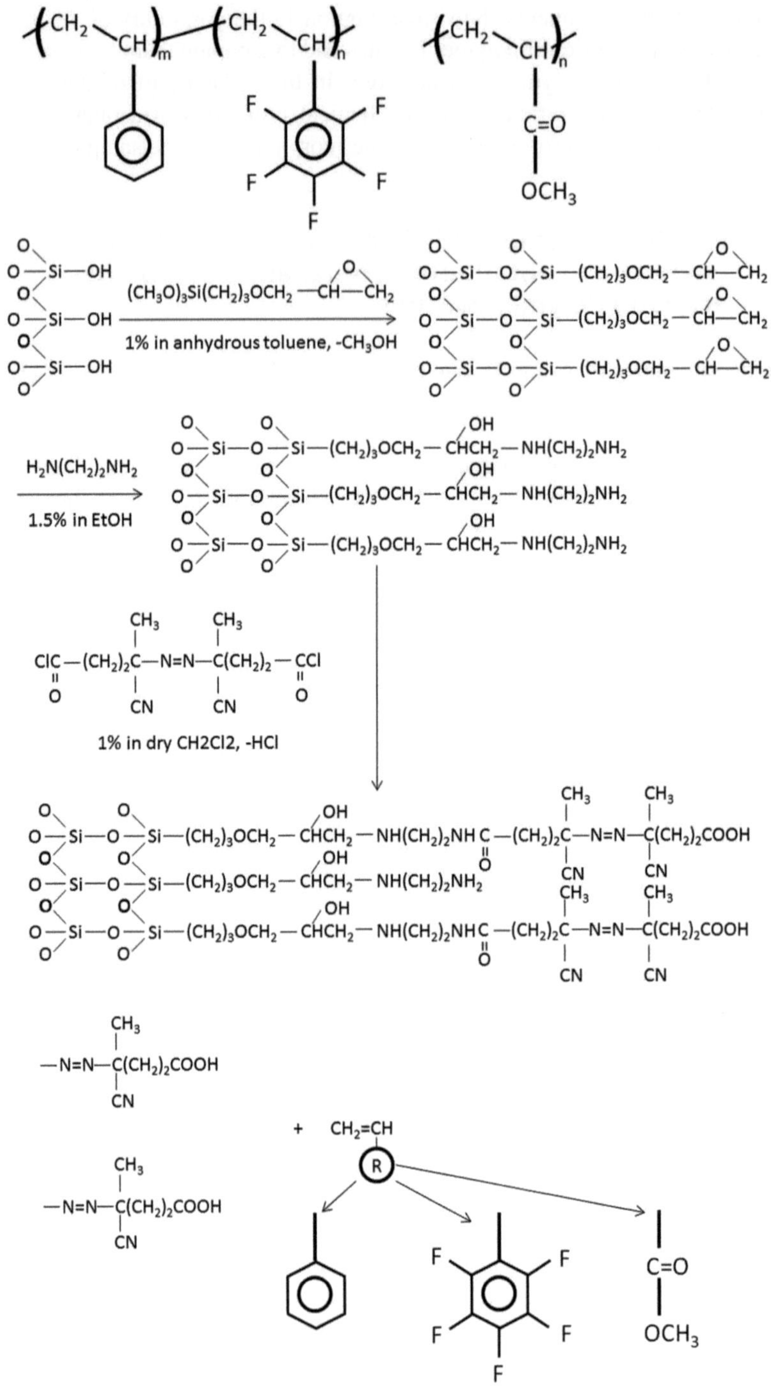

Fig. 4.17 (*Top*) Structure formulas of PSF (*left*) and PMA (*right*). (*Bottom*) Schematic representation of the procedure for brush fabrication: modification of Si wafer with GPS, reaction with

between 20 and 50 nm and elastic constants ranging from 0.01 to 50 N/m (first article) and with radii between 10 and 30 nm and elastic constants ranging from 1 to 30 N/m (second article).

Spring constants were determined through a method combining finite element analysis (FEA) calculations and resonance frequency measurements [17]. Tip radii were evaluated through scanning of reference gold nanoparticle samples.

Force–volume measurements at different temperatures were performed in the DI thermal stage sample holder. Commonly, 64 × 64 curves on 1 μm^2 surface areas were acquired.

In the first article [12], the glass transition temperature was determined through DSC measurements and scanning thermal microscopy (SThM) with microthermal analysis (μTA). In this method, the glass transition temperature is determined from heat dissipation data [18].

Elastic Modulus of PSF and PMA Polymer Brushes The analysis of the mechanical properties of polymer brushes starts in the first article [12] with the characterization of the layer composition. Via Tapping Mode measurements, the authors could assess the homogeneity of the samples.

Two properties of the brush layers are studied through force–volume measurements: the adhesion force and the elastic modulus.

Figure 4.18 shows two representative force–distance curves on PSF and PMA layers (left) and the corresponding deformation–force curves (right).

The two force–distance curves in the picture show the two basic differences between both polymers: the stiffness of PSF is considerably larger than that of PMA and the adhesion of PSF is noticeably smaller than that of PMA.

The stiffness, defined as $d\delta/dZ$, varies from 0.6 to 0.8 for PSF and from 0.2 to 0.4 for PMA. The larger compliance of the PMA layer is confirmed by the deformation–force curves, showing that, for the same load, the penetration of the AFM tip into the PMA layer is more than five times higher than into the PSF layer.

Histograms of the normalised pull-off force F/R (left) and of the elastic modulus (right) are shown in Fig. 4.19 both for PMA (grey) and PSF layers (black). Elastic moduli were calculated in all measurements applying Hertz theory. The elastic modulus of PSF is between 1 and 1.2 GPa and that of PMA between 50 and 60 MPa. The PSF value is close to typical values measured for glassy polymers; the PMA value is typical for rubbery polymers. This agrees with the fact that the glass transition temperature of PSF is ca. 108 °C and that of PMA is about 5 °C.

Under the loading conditions exploited in this work, the contact radius estimated through Hertz equation does not exceed 9 nm. Since the gyration radius is 18 nm for PSF and 15 nm for PMA, it can be concluded that PSF and PMA brushes are homogeneous materials with elastic properties typical for glassy and rubbery bulk polymers even at a scale finer than one macromolecular chain.

Fig. 4.17 (continued) ethylendiamine, attachment of Cl-ABCPA, and "grafting-from" growth. Adapted with permission from [12]. Copyright 2003. American Chemical Society

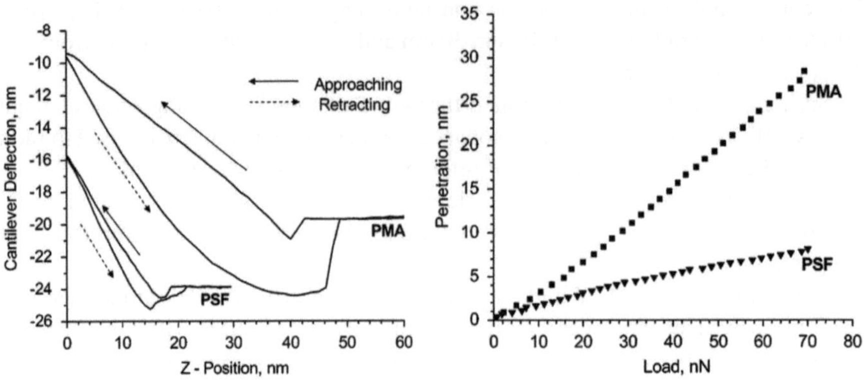

Fig. 4.18 (*Left*) Representative force–distance curves for PSF and PMA layers. Curves have been shifted for clarity. (*Right*) Deformation–force curves for PMA and PSF layers. Reprinted with permission from [12]. Copyright 2003. American Chemical Society

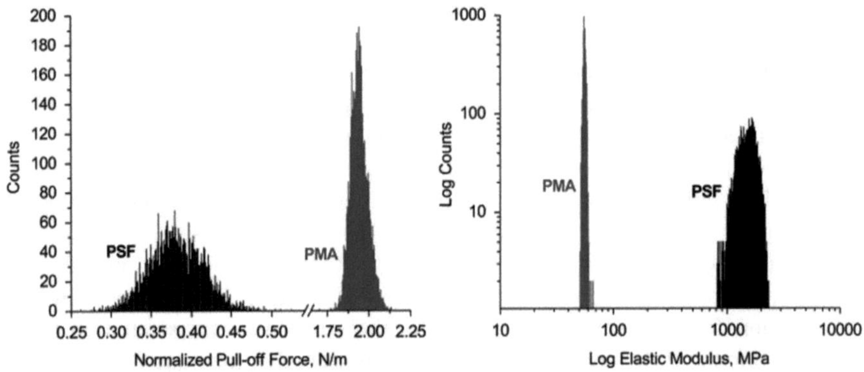

Fig. 4.19 Histograms of the normalised pull-off force F/R (*left*) and of the elastic modulus (*right*) for PMA (*grey*) and PSF layers (*black*). The histograms are calculated from a 64×64 force–volume measurement. Reprinted with permission from [12]. Copyright 2003. American Chemical Society

Measurements on the brush layers have been compared with measurements on spin-coated films of the same polymers, yielding elastic moduli between 1.2 and 1.6 GPa for PSF and between 7 and 10 MPa for PMA.

The adhesive force is much higher for PMA than for PSF (1.95 N/m compared to 0.37 N/m). The difference in the adhesion of the two brushes results not only from the fact that PSF is glassy and PMA rubbery, but also from the presence of fluorinated groups in PSF opposite to the non-fluorinated surface of PMA. Due to the fluorinated groups, PSF has a substantially lower surface energy than PMA, on the surface of which polar double bonds are present.

The mechanical properties of the PSF brushes were studied also in dependence of temperature, both with force–volume and microthermal analysis (μTA).

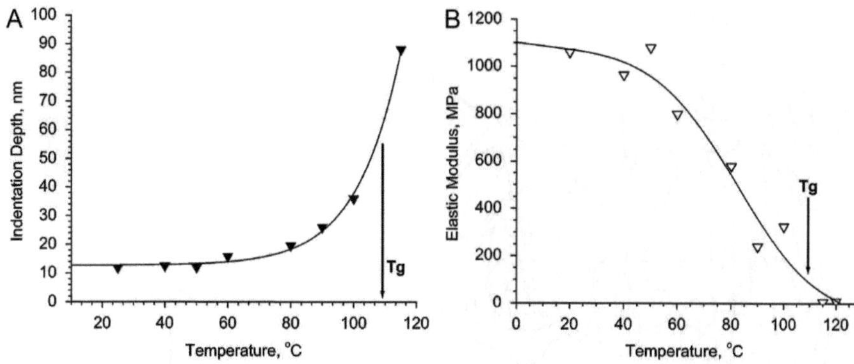

Fig. 4.20 (*Left*) Penetration depth into the PSF layer with the same load as a function of temperature. (*Right*) Elastic modulus of the PSF layer as a function of temperature. Reprinted with permission from [12]. Copyright 2003. American Chemical Society

μTA measurements of heat dissipation allowed direct evaluation of the glass transition temperature for the PSF brush layer, whereas PMA was not measured because technical constraints limit the temperature range to those above room temperature and the glass transition temperature of PMA is about 5 °C. The glass transition temperature of PSF determined from the μTA measurement was 109 °C, close to the glass transition temperature obtained from a DSC experiment on a bulk PSF sample (108 °C).

Figure 4.20A shows the penetration depth into the PSF layer at constant load, but with increasing temperature T. The indentation is nearly constant for $T < 60$ °C and increases sharply for $T > 90$ °C. The elastic modulus, calculated from force–volume measurements at each temperature and shown in Fig. 4.20B, is nearly constant for $T < 60$ °C and decreases to ca. 15 MPa for $T > 110$ °C, as expected for a polymer with $T_g = 109$ °C (see Sects. 1.6, 3.10 and 3.11).

In the second article [13], both polymers are present on one substrate as a binary brush. Immersing the sample in a selective solvent, i.e. toluene or acetone, for 5 min, it is possible to switch the binary brush.

As shown schematically in Fig. 4.21, acetone, a selective solvent for PMA, leads to the collapse of PSF, forming clusters, and to the swelling of PMA, segregating at the top end of the brush. The swelling of PMA is enhanced by its low glass transition temperature and by a favourable Flory interaction parameter. The resulting sample consists of a 110–120 nm thick PMA brush layer and a 10–15 nm thick PSF layer, completely separated.

In contrast, toluene is a selective solvent for PSF. Hence, the collapsing polymer, forming clusters, is PMA, whereas PSF swells and segregates at the top end of the brush. Yet, due to its high glass transition temperature, limiting the mobility of PSF chains, and to an unfavourable Flory interaction parameter, the swelling of PSF is much more moderate than that of PMA. As a consequence, in this case the two layers, having comparable thickness (30–40 nm for PMA and 50 nm for PSF), are not completely separated. Topography images acquired in Tapping Mode after

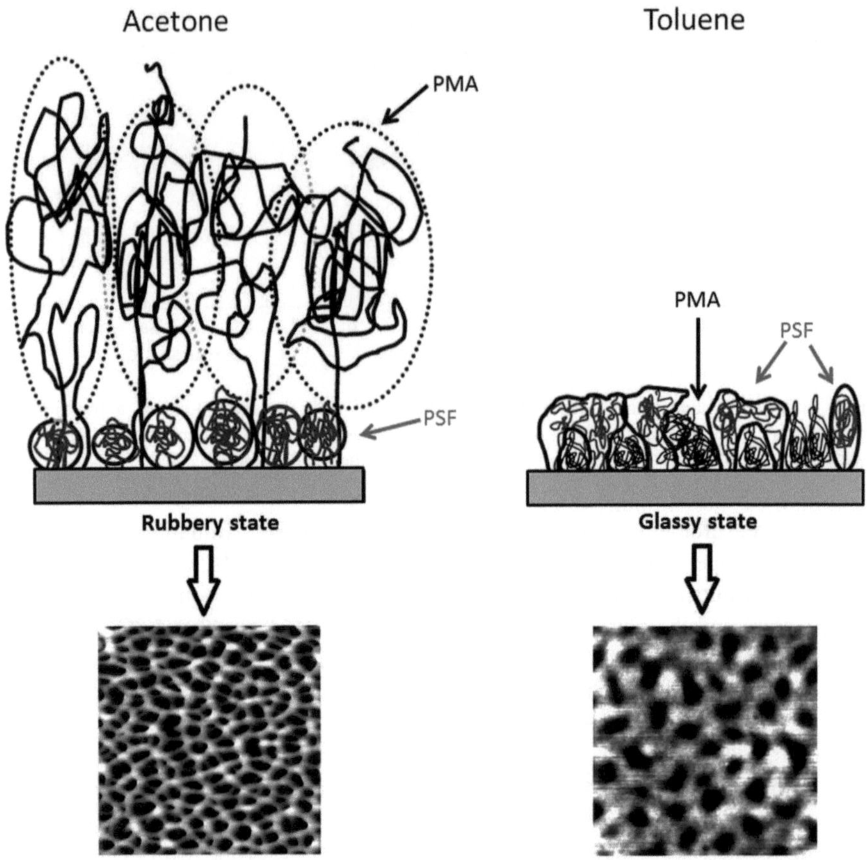

Fig. 4.21 Schematic description of the reordering of PSF and PMA chains due to exposure to selective solvents, i.e. acetone (*left part*) and toluene (*right part*). In the bottom part, AFM topography images acquired in Tapping Mode for both configurations. Both pictures are 1 μm^2 large; the Z scale is 150 nm for the *left picture* and 10 nm for the *right* one. Adapted with permission from [13]. Copyright 2003. American Chemical Society

exposure to the two solvents are shown in the bottom part of Fig. 4.21. Both images are 1 μm^2 large, but the Z scales are 150 nm for the left image and 10 nm for the right one. The height difference between "tops" and "holes" is about 100 nm in the left image and about 10 nm in the right one. The overall thickness of the samples has been measured with Tapping Mode topographies over a scratch.

The morphology of the sample and in particular the switching of the binary polymer brush due to selective solvent are confirmed by measurements of the mechanical properties. Such measurements revealed to be rather difficult, because of several factors:

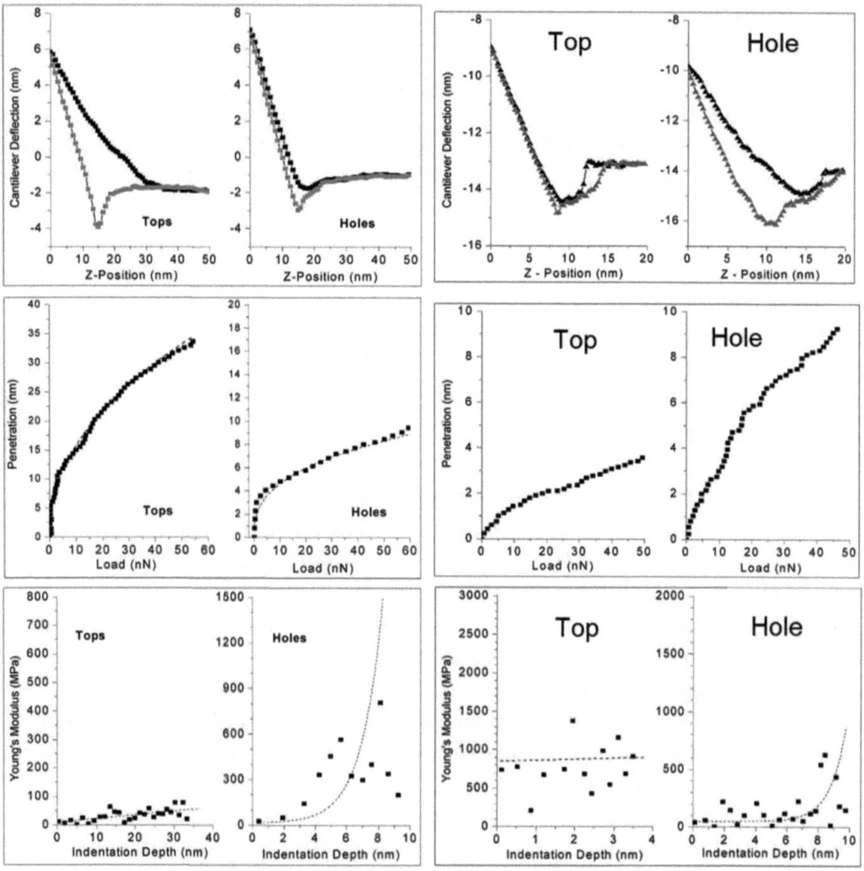

Fig. 4.22 Representative force–distance curves (*top*, *black symbols* for approach curves, *grey symbols* for retraction curves), deformation–load curves (*middle*), and elastic modulus as a function of deformation (*bottom*) for the binary brushes exposed to acetone (*left*) and toluene (*right*). For each sample, measurements have been performed on "tops" location and in holes, as indicated in the figure. Adapted with permission from [13]. Copyright 2003. American Chemical Society

1. Due to the dimensions of the nanodomains, high lateral resolution is required. In turn, this requires a very small tip radius.
2. Because of repeated calibration of the sensitivity, implying indentation on silicon, and repeated indentation of the polymer layers, the tip is likely to become blunt and contaminated.
3. Since PSF and PMA have very different mechanical properties, measurements had to be repeated with cantilever of very different stiffness.

Figure 4.22 shows results of force–volume measurements on top and holes of the binary brush after exposure to acetone (left) and toluene (right).

Accordingly to the switching of the composition of the topmost layer just described, when the sample has been exposed to acetone, the "holes" consist of PSF and the "tops" of PMA.

Considering the deformation–load curves shown in the middle row, it can be seen that the penetration into the material on the tops is considerably higher than the penetration into the material in the holes for the same load (ca. 30 vs. 4 nm). In particular, the deformation–load curve for the material in the holes is very steep at the beginning of the loading process. The different behaviour of the sample at the two locations is evident also considering the respective force–distance curves shown in the top row.

The elastic modulus of the sample, calculated at different penetrations, shown in the last row, indicates that the polymer on the tops is in a rubbery state. The modulus varies with the depth between 25 MPa at 10 nm penetration and 70 MPa at 40 nm penetration. The increase of Young's modulus is due to the effect of the substrate.

In the holes, the elastic modulus steeply increases up to about 1 GPa after 4 nm indentation. This dependence of the elastic modulus on indentation, explaining the steep rise of the deformation at small loads, indicates that a thin PMA layer is present on top of the PSF brushes in the holes.

Turning to the measurements on the sample exposed to toluene (right part of Fig. 4.22), it can be seen that the force–distance curve and the deformation–load curve on the "tops" resemble those on the holes of the sample exposed to acetone. But in this case, the PSF is not covered with a thin layer of PMA; hence, there is no hysteresis in the contact part of the force–distance curve and no steep part at the beginning of the deformation–load curve.

The elastic modulus measured on the top locations has a constant value of about 1 GPa. In the holes, the initial modulus is about 50 MPa up to a penetration of about 10 nm. At this deformation, the elastic modulus rises steeply. The increase is due again to the effect of the substrate.

In both measurements, the values of the elastic moduli of both components are next to those obtained on bulk samples of the same polymers, i.e. 6–50 MPa for PMA and 0.8–1.2 GPa for PSF.

References

1. Schubert DW (1997) Spin coating as a method for polymer molecular weight determination. Polym Bull 38:177–184
2. Schubert DW, Dunkel T (2003) Spin coating from a molecular point of view: its concentration regimes, influence of molar mass and distribution. Mater Res Innov 7:314–321
3. Opdahl A, Somorjai GA (2001) Stretched polymer surfaces: atomic force microscopy measurement of the surface deformation and surface elastic properties of stretched polyethylene. J Polym Sci B 39:2263–2274
4. Cappella B, Silbernagl D (2007) Nanomechanical properties of mechanical double-layers: a novel semiempirical analysis. Langmuir 23:10779–10787

5. Silbernagl D, Cappella B (2010) Mechanical properties of thin polymer films on stiff substrates. Scanning 32:282–293
6. Silbernagl D, Cappella B (2009) Reconstruction of a hidden topography by single AFM force–distance curves. Surf Sci 603:2363–2369
7. Krämer G, Griepentrog M, Bonaccurso E, Cappella B (2014) Study of morphology and mechanical properties of polystyrene–polybutadiene blends with nanometer resolution using AFM and force–distance curves. Eur Polym J 55:123–134
8. Hutter JL, Bechhoefer J (1993) Calibration of atomic-force microscope tips. Rev Sci Instrum 64:1868–1878
9. Cappella B (2011) Mechanical properties and adhesion of a micro structured polymer blend. Polymers 3:1091–1106
10. Cappella B, Kaliappan SK, Sturm H (2005) Using AFM force–distance curves to study the glass-to-rubber transition of amorphous polymers and their elastic-plastic properties as a function of temperature. Macromolecules 38:1874–1881
11. Hayashida K, Tanaka H, Watanabe O (2009) Miscible blends of poly(butyl methacrylate) densely grafted on fumed silica with poly(vinyl chloride). Polymer 50:6228–6234
12. Lemieux M, Minko S, Usov D, Stamm M, Tsukruk VV (2003) Direct measurement of thermoelastic properties of glassy and rubbery polymer brush nanolayers grown by "grafting-from" approach. Langmuir 19:6126–6134
13. Lemieux M, Usov D, Minko S, Stamm M, Shulha H, Tsukruk VV (2003) Reorganization of binary polymer brushes: reversible switching of surface microstructures and nanomechanical properties. Macromolecules 36:7244–7255
14. Zhao B, Brittain WJ (2000) Polymer brushes: surface-immobilized macromolecules. Prog Polym Sci 25:677–710
15. Luzinov I, Julthongpiput D, Liebmann-Vinson A, Cregger T, Foster MD, Tsukruk VV (2000) Epoxy-terminated self-assembled monolayers: molecular glues for polymer layers. Langmuir 16:504–516
16. Tsukruk VV, Luzinov I, Julthongpiput D (1999) Sticky molecular surfaces: epoxysilane self-assembled monolayers. Langmuir 15:3029–3032
17. Hazel JL, Tsukruk VV (1999) Spring constants of composite ceramic/gold cantilevers for scanning probe microscopy. Thin Solid Films 339:249–257
18. Tsukruk VV, Gorbunov VV, Fuchigami N (2003) Microthermal analysis of polymeric materials. Thermochim Acta 395:151–158

5. Silbernagl D, Cappella B (2010) Mechanical properties of thin polymer films on stiff substrates. Scanning 32:282–293
6. Silbernagl D, Cappella B (2009) Reconstruction of a hidden topography by single AFM force-distance curves. Surf Sci 603(2):[illegible]
7. Kumar B[illegible], [illegible] M, Bonaccurso E, Cappella B (2014) Study of morphology and mechanical properties of polystyrene–polybutadiene blends with nanometre resolution using AFM and force–distance curves. Eur Polym J 55:[illegible]
8. [illegible]

Polymer Blends

5

Abstract

Polymer blends are together with polymer thin films on stiff substrate a broad category of inhomogeneous samples, whose local properties can be advantageously investigated with an AFM.

The present chapter about polymer blends is divided into two parts. In the first part, experiments on model blends and confined polymers are reviewed, whereas the second part deals with the characterisation of microstructured blends.

5.1 Model Blends and Confined Polymers

The samples examined in the two following hands-on examples, a model blend [1] and a confined polymer [2], give the possibility to characterise the mechanical properties of samples consisting of two phases, but with a simple morphology. In both samples the two phases have macroscopic dimensions and there is only one interface in between.

In the first experiment [1], the two phases are two immiscible polymers, poly(*n*-butyl methacrylate) and polystyrene. The sample preparation, aimed to create a common interface without any topography difference, influences strongly the mechanical properties of both phases. Nevertheless, it can be shown that the mechanical properties of both components can be determined as a function of temperature in agreement with other AFM measurements and with measurements performed with other techniques; differences between the measurements can be explained through the sample preparation; the two phases can be distinguished with sub-micrometre resolution in maps of the elastic modulus; and the local mechanical properties at the interface can be characterised and explained in detail.

In the second reported experiment [2], the sample preparation, even if considerably more complicated, does not influence noticeably the mechanical properties of the examined polymer, poly(methyl methacrylate). Hence, interface effects on the mechanical properties of the polymer can be analysed more precisely.

B. Cappella, *Mechanical Properties of Polymers Measured through AFM Force-Distance Curves*, Springer Laboratory, DOI 10.1007/978-3-319-29459-9_5

5.2 Hands-on Example 17: Spatial Variation of the Thermomechanical Properties of a Model Polystyrene/Poly(*n*-butyl methacrylate) Blend

This hands-on example deals with the spatial variation of the thermomechanical properties of a model blend made of poly(*n*-butyl methacrylate) (PnBMA) and polystyrene (PS). Measurements have been carried out in the group of Cappella [1].

Sample Preparation and Instrumentation PnBMA and PS were purchased from Scientific Polymer Products Inc. (Ontario, NY) and from BDH Chemicals Ltd. (Poole, England), respectively. The molecular weight, the polydispersity index and the glass transition temperature, provided by the suppliers, are $M_w = 319$ kDa, $M_w/M_n \leq 2.58$ and $T_g = 22$ °C for PnBMA and $M_w = 100$ kDa, $M_w/M_n \leq 2$ and $T_g = 100$ °C for PS.

The sample preparation of the model blend plays a very important role, since it influences strongly the mechanical properties of the sample, as shown by the measurements. The blend was prepared in two steps.

First Step Homogeneous films of both polymers were prepared by melting PnBMA and PS between glass slides, in vacuum at 140 °C and 200 °C, respectively. Heating was provided through a 340-temperature controller (Lake Shore Cryotronics, Westerville, OH). Pressure was applied on the glass slides during the melting with a dual spring, to the aim of obtaining uniformly thick films with flat surfaces. Subsequently, the homogeneous films have been cut in two square films, 1 cm^2 large.

Before the second step of the preparation, the edges of the PS film were imaged in tapping mode, in order to determine the angle of cut, which was $20 \pm 5°$.

Second Step The PnBMA film, placed on top of the PS film, was molten again at 140 °C between two glass slides. Since pressure was applied again with a dual spring, PnBMA flowed down and surrounded the PS film.

The bottom side of the sample, in contact with the heater, was used for measurements.

The PS–PnBMA model blend was approximately 200 μm thick; hence, even large indentations were not affected by the substrate.

AFM force–displacement curves were acquired using a MFP3D microscope (Asylum Research, Santa Barbara, CA) equipped with Pointprobe NCL cantilevers (Nanosensors, Germany) with a spring constant $k_c = 45$ N/m. The spring constant was measured through the method of Hutter and Bechhoefer [3] (see Sect. 2.2.1).

For measurements at different temperatures, a metallic disc at the base of the polymer was heated using a 340-temperature controller. The surface temperature of the sample was detected using a PT100 fixed directly on the polymer surface.

The surface temperature was equilibrated overnight; the temperature stayed constant for several days, with a maximum variation of 0.3 °C on the surface.

Force–volume measurements were performed at 32, 38.5, 45, 51.4, 57.4, 63.6 and 70.1 °C. The acquisition frequency of the force–displacement curves was 1 Hz in all measurements. At each temperature, three force–volume measurements were performed: two force–volumes with 10 × 10 curves on 80 × 80 μm^2 large areas at a distance of about 2 mm from the interface, both on PnBMA and PS, and one force–volume with 100 × 100 curves on a 80 × 80 μm^2 large area across the interface. All force–displacement curves had a maximum cantilever deflection of 400 nm, corresponding to a maximum applied force of 18 μN.

Measurements away from the interface were repeated on another PS–PnBMA model blend (sample II), in order to show the repeatability of the measurements. In this case the temperatures were 28.9, 34.3, 40.3, 45.4, 51.2, 57 and 63 °C.

Thermomechanical Properties of a Model PS–PnBMA Blend The first group of results of the present experiment concerns the mechanical properties of both polymers in the blend away from the interface or, in other words, the bulk values of the elastic moduli of both constituents of the blend.

The elastic moduli of the polymers were calculated through the hyperbolic fit (see Sect. 1.8.3) as $E = \frac{3(1-\nu^2)}{4}\frac{k_c}{\sqrt{R}}\frac{1}{\gamma}$. In Sect. 3.14 it has been shown that differences between the instantaneous elastic modulus E_0 and Young's modulus E are negligible.

Figure 5.1 shows the logarithm of Young's modulus versus the temperature obtained in the following measurements:

1. Present measurement on the first model blend (sample I), PnBMA (empty circles) and PS (empty squares) bulk values.
2. Present measurement on the second model blend (sample II), only PnBMA bulk values (empty triangles).
3. AFM measurement on a homogeneous film in [4], PnBMA (filled circles) (see Sect. 3.14).
4. AFM measurement on a homogeneous film in [5], PS (filled squares) (see Sect. 3.15).
5. DMA measurement on a homogeneous film in [4], PnBMA (solid line) (see Sect. 3.14).

The plot shows a very good agreement between the present AFM measurements, other AFM measurements and the DMA data. Such a good agreement is remarkable when two basic differences between the measurements are considered. First of all, the four AFM measurements (two model blends and two homogeneous films) have been performed with four different cantilevers and different tips with different radii. Second, the homogeneous films have been cast from concentrated polymer solutions, whereas the blend has been prepared by melting the polymers in vacuum between two glass plates under pressure. The significant agreement between the

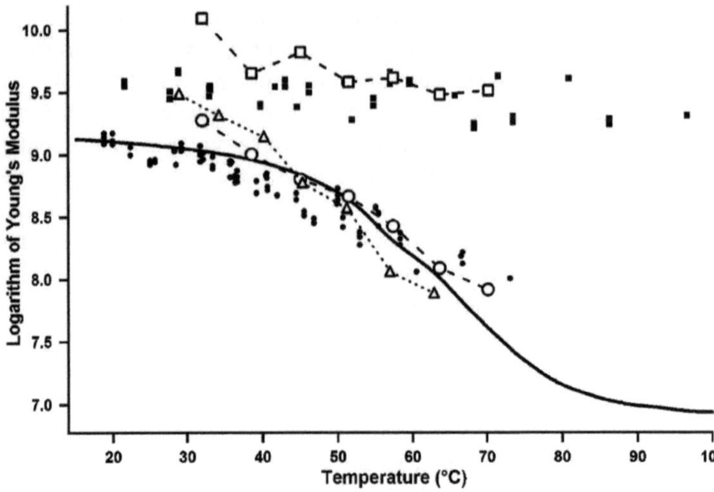

Fig. 5.1 Logarithm of the elastic modulus E versus the temperature for PnBMA (*empty circles*, sample I, and *empty triangles*, sample II) and PS (*empty squares*) away from the interface, compared with log(E) of PnBMA (*filled circles*) and PS (*filled squares*) determined in [4] and [5], respectively (see Sects. 3.14 and 3.15). Also log(E) of PnBMA measured through DMA in [4] is shown as a *thick continuous line* (see Sect. 3.14). Reprinted with permission from [1]. Copyright 2006. American Chemical Society

values of the elastic moduli of PnBMA and PS obtained in different experiments is a proof of the repeatability of the measurements and of the accuracy of the analysis.

Nevertheless, the measurements on the blend present a fundamental discrepancy, when compared with the measurements on the homogeneous samples.

At low temperatures the elastic modulus of the homogeneous PnBMA is lower than bulk Young's modulus of PnBMA in the blend, both on sample I (empty circles, only at 32 °C) and on sample II (empty triangles, up to 40.3 °C). Starting from 38.5 °C for sample I and 45.4 °C for sample II, the bulk elastic modulus of PnBMA in the blend is the same as that of homogeneous PnBMA. For PS, bulk Young's modulus of the polymer in the blend is larger than that of the homogeneous films at every temperature.

This discrepancy is due to the sample preparation. During the preparation of the model blend, PS and PnBMA films were molten and cooled down under pressure. Hence, internal stresses have been engendered within the films. The polymers can relax such residual stresses at temperatures above their T_g, but all experimental temperatures are well below the glass transition temperature of PS (100 °C).

As already shown in Sects. 3.14 and 3.15, the elastic modulus of PnBMA decreases by more than one order of magnitude between 32 and 70 °C. By contrast, the elastic modulus of PS does not decrease significantly.

The actual goal of measurements on a model blend is to characterise the mechanical properties of the polymers next to the interface.

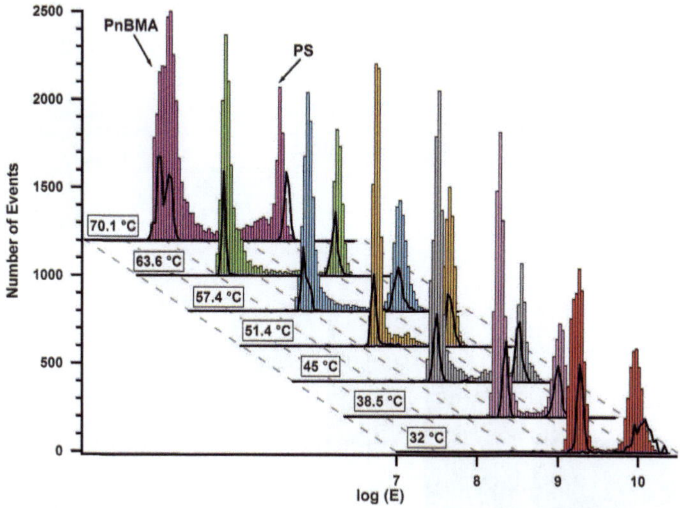

Fig. 5.2 Histograms of log(E) for the measurements across the PS–PnBMA interface (*bars*) and log(E) of both polymers away from the interface (*solid lines*). The histograms have been shifted both horizontally and vertically for clarity. The peak on the *left* (*right*)-*hand side* corresponds to log(E) of PnBMA (PS). Reprinted with permission from [1]. Copyright 2006. American Chemical Society

The elastic modulus of the PS–PnBMA model blend has been calculated from the fit of the $D^{3/2}$ curves also for force–volume measurements across the interface. Figure 5.2 shows the histograms of log(E) obtained from the measurements across the interface (bars) together with histograms of the bulk values of log(E) of both polymers (solid lines).

The histograms have been shifted both horizontally and vertically for clarity. The peak at lower (larger) values corresponds to PnBMA (PS). Again, the elastic modulus of PS varies very little in comparison to that of PnBMA.

A first important result, which can be observed in Fig. 5.2, is that the histograms of the bulk values overlap the ones obtained over the interface. A more detailed analysis reveals that the histograms obtained from PS close to the interface can be fitted with a Gaussian function, whereas the histograms obtained from PnBMA close to the interface have a shoulder on the right-hand side. This means that Young's modulus of PnBMA close to the interface is higher than the bulk value.

Maps of the elastic modulus permit to relate the mechanical properties to the morphology of the blend.

Figure 5.3 shows grey scale maps of the logarithm of the elastic modulus obtained at different temperatures and at different locations across the interface. All images are shown with the same grey scale, in which black corresponds to the minimum measured elastic modulus.

The PnBMA regions become darker with increasing temperature as a result of the decrease in Young's modulus. By contrast, the parts of the images on PS do not

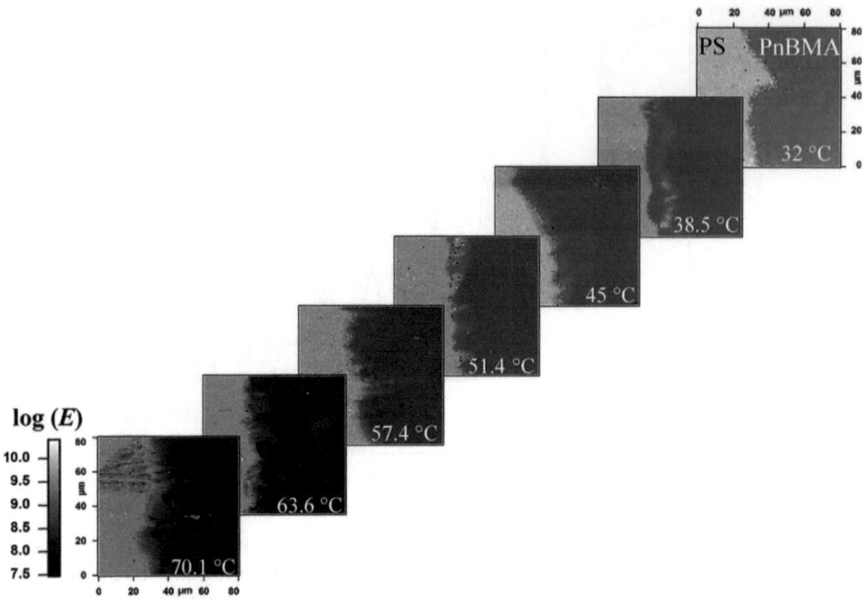

Fig. 5.3 Maps of log(E) obtained from the measurements across the PS–PnBMA interface at various temperatures as indicated in the images. All images have the same grey scale. Reprinted with permission from [1]. Copyright 2006. American Chemical Society

show significant changes. The transition between the moduli of the two polymers is rather sharp and permits to localise the interface with a resolution of ca. 800 nm.

The interfacial region can be observed more in detail in Fig. 5.4, showing the averaged line profiles of log(E) across the interfacial region at all temperatures versus the distance from the interface x_I.

Once again, it can be noted that the plateau values corresponding to the modulus of PS are bunched together on the left-hand side of the graph for all temperatures, whereas the modulus of PnBMA decreases with increasing temperature.

At each temperature, the elastic modulus of PS is rather constant until the interface is reached, whereas Young's modulus of PnBMA depends strongly on x_I. In particular, with increasing x_I, the modulus decreases from the bulk value of PS to the bulk value of PnBMA, indicated by the arrows on the right-hand side of the image. The width of the region, in which the elastic modulus of PnBMA varies with x_I, increases with increasing temperature, too. At 70.1 °C the transition region between the bulk values of PS and PnBMA is larger than 80 μm, and it was necessary to acquire a second force–volume on an $80 \times 80\ \mu m^2$ area adjacent to the first one. The second line profile, not shown for clarity, matches the PnBMA bulk value of log(E).

The transition region introduced here is several micrometres wide and is defined through the gradient of Young's modulus of the sample; hence, it is not the transition region commonly related to the interface between two polymers, which

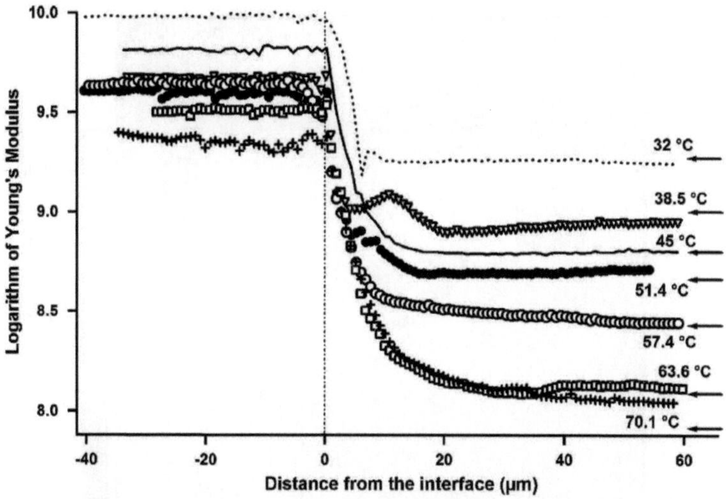

Fig. 5.4 Averaged line profiles of log(*E*) across the interface at all temperatures versus the distance from the interface x_I. The *arrows* on the *right-hand side* indicate the bulk PnBMA value of log(*E*). Reprinted with permission from [1]. Copyright 2006. American Chemical Society

is of the order of some nanometres for immiscible polymers such as PS and PnMBA and is defined primarily by the gradient of the composition of the sample.

The dependence of the elastic modulus of PnMBA on x_I gives the possibility of characterising the morphology of the sample through its Young's modulus. Figure 5.5a shows the contours of log(*E*) = 9.48 (dark red line) down to log(*E*) = 9 (light yellow line), in steps of 0.08 (lines of increasing brightness, very close to each other). The contours of log(*E*) = 8.6 (green line) and log(*E*) = 8.5 (blue line) are shown, too. A contour is a curve joining the points at which log(*E*) goes through a certain value.

The contours are drawn on the topography image across the interface (image in grey scale) at 57.4 °C. In this image, the two phases can be distinguished only because it was acquired after the force–volume, and plastic deformations engendered by the force–distance curves left small cavities or imprints on the blend surface. Such deformations could be relaxed only on the PnBMA surface and not on the PS surface. Without such imprints, it would be impossible to distinguish the two phases in a topography image.

The two phases can indeed be distinguished through the contours of log(*E*), since the contours at values between 9.48 and 9 follow the interface. The contours of log(*E*) = 8.6 and log(*E*) = 8.5 depart from the interface, since these two values are next to the PnBMA bulk value of log(*E*) (8.42).

Figure 5.5b shows the line profiles of the topography and of log(*E*) at 57.4 °C at the same position. The topography does not present any discontinuity at the interface (dotted line), which can be localised only via the profile of the elastic

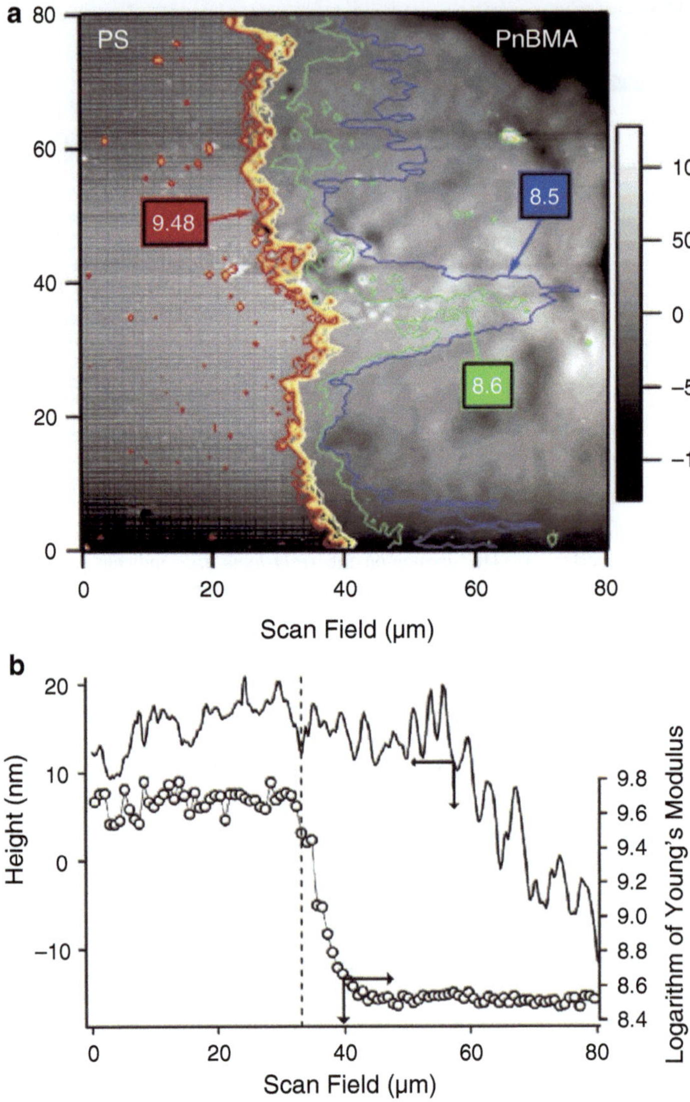

Fig. 5.5 (**a**) Topography image (*grey scale*) of the model blend across the interface at 57.4 °C and superimposed contours of log(E) = 9.48 (*dark red line*) down to log(E) = 9 (*light yellow line*), in steps of 0.08 (*lines of increasing brightness*, very close to each other). Also the contours of log(E) = 8.6 (*green line*) and of log(E) = 8.5 (*blue line*) are shown. (**b**) Line profile of the topography (referred to left axis) and of log(E) (referred to the right axis) of the model blend across the interface. The *dotted line* represents the position of the interface. Reprinted with permission from [1]. Copyright 2006. American Chemical Society

modulus. It is evident that the topography does not influence the measured elastic modulus.

As already pointed out, the stiffening of PnBMA in a region at the interface being several micrometres wide cannot be explained through the existence of the interfacial region, which is only some nanometres wide. Rather, it must be related to geometrical constraints of the PnBMA phase.

One of the causes for the stiffening of PnBMA at the interface is the presence of macroscopic stresses developed in PnBMA, which are also the reason for the stiffening of PnBMA away from the interface at low temperatures (see Fig. 5.1). Since PnBMA and PS are immiscible, the PnBMA phase tries to separate from the PS phase by dewetting, in order to reduce its surface free energy (see Sect. 5.5).

Compared to PS, PnBMA chains have a larger free volume, as PnBMA is above its T_g. Hence, PnBMA chains are more mobile and PnBMA has the possibility to increase its volume due to thermal expansion. The thermal expansion can occur away from the interface, if the temperature is high enough, because PnBMA is not subjected to any geometrical constraint. Yet, at the interface, the thermal expansion is hindered by the presence of PS, which is a comparatively stiff polymer and cannot mix with PnBMA. In other words PS, due to its immiscibility with PnBMA and to its relatively high stiffness at all experimental temperatures, plays the role of a hard wall and represents a barrier for the thermal expansion of PnBMA. The thermal expansion of PnBMA and its tendency to dewet the PS phase, together with the presence of a barrier at the interfacial region, engenders stresses in the PnBMA phase.

The hypothesis of stresses induced by the thermal expansion is confirmed by severe changes of the sample topography above 70.1 °C, accompanied by massive stresses. At this temperature PnBMA had enough thermal energy to separate from PS, and a groove was formed at the interface with a width of about 30 μm and a depth of more than 6 μm; parallel to the grove, a pile with a height of several micrometres was generated.

Another conceivable reason for the stiffening of PnBMA at the interface is the influence of underlying PS, acting as the substrate of a mechanical double layer (see Sects. 1.10 and 4.1–4.5).

As already said, the angle at the edge of the PS film, measured in Tapping Mode, is $20 \pm 5°$. Hence, since the lateral distance between two adjacent curves in the force–volume is 800 nm, already after one point from the interface, coinciding with the PS edge, the thickness of PnBMA on top of PS is about 300 nm. It is important to remember that the values of the elastic modulus are calculated from the instantaneous region of the $D^{3/2}$ curve, i.e. from the parameter γ (see Sects. 1.8.3 and 3.14). The maximum elastic deformation of PnBMA at 57.4 °C and at some micrometres from the interface is about 60 nm. This means that the thickness of the PnBMA film on top of PS is much larger than the deformation. Consequently, the mechanical properties of PnBMA probed by the cantilever are not influenced by the underlying PS (compare Sect. 4.2).

Summarising, it can be assessed that the effect of the underlying PS contributes to the stiffening of PnBMA next to the interface, but it cannot be the sole reason; in

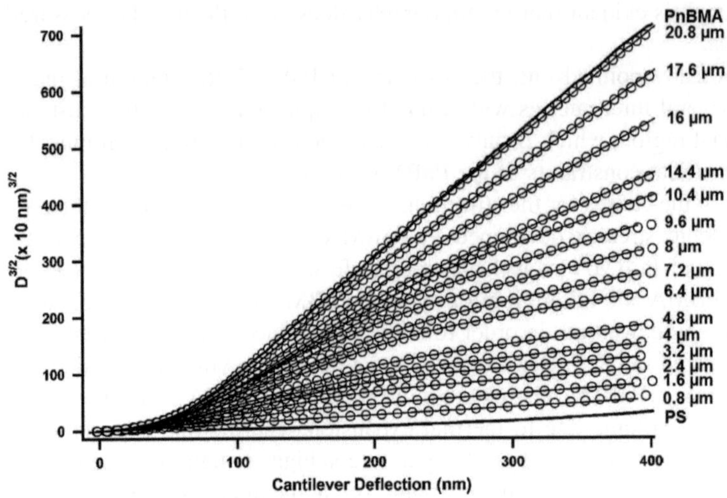

Fig. 5.6 $D^{3/2}$ curves (*empty circles*) acquired at 57.4 °C on PnBMA at increasing distance from the interface x_I, as indicated on the *right side* of the curves. The $D^{3/2}$ curves on PS and on PnBMA far from the interface are shown as *thick continuous lines*. The curve acquired at $x_I = 21.6$ μm overlaps the curve on PnBMA away from the interface. The *continuous lines* represent the fit with Eq. (5.1). Reprinted with permission from [1]. Copyright 2006. American Chemical Society

fact, the hard-wall effect at the interface plays a major role in the stiffening of PnBMA.

The detailed analysis of the $D^{3/2}$ curves on PnBMA close to the interface confirms this assumption. Figure 5.6 shows $D^{3/2}$ curves acquired at 57.4 °C on PnBMA at increasing distance from the interface x_I, as indicated in the figure, together with the $D^{3/2}$ curves on PS and on PnBMA far from the interface.

It can be observed that:

1. The first linear region of the curves ($D^{3/2} \approx \gamma\delta$) has a higher slope, i.e. the sample is more compliant, with increasing x_I, but it does not deviate from the linear behaviour. Hence, the instantaneous region of the curves is not influenced by the underlying PS, since the ratio D/t_{PnBMA} (with t_{PnBMA} thickness of the PnBMA phase on top of PS) is very small, even for very small x_I. The dependence of the sample stiffness on x_I is engendered by stresses due to the hard-wall effect caused by the presence of the PS phase.
2. Starting at a certain deformation, the second linear region ($D^{3/2} \approx (\beta + \alpha)\delta$) deviates from the linear behaviour, and deviations become smaller and smaller with increasing x_I. These deviations are engendered by the underlying PS, since D/t_{PnBMA} is larger due to larger deformations.
3. With increasing x_I the curves gradually go from a typical PS curve to a typical PnBMA curve, with the curve acquired at $x_I = 21.6$ μm (not shown) overlapping the curve on PnBMA far away from the interface.

Due to their shape, the curves at the interface cannot be fitted with a hyperbola (see Sect. 1.8.3). For their fit, following equation has been used:

$$D^{3/2} = (\beta_1\delta_c - \varepsilon_1) + \sqrt{\alpha_1^2\delta_c^2 - 2\varepsilon_1(\beta_1 - \gamma_1)\delta_c + \varepsilon_1^2} - \left[(\beta_2\delta_c - \varepsilon_2) + \sqrt{\alpha_2^2\delta_c^2 - 2\varepsilon_2(\beta_2 - \gamma_2)\delta_c + \varepsilon_2^2}\right] \tag{5.1}$$

The first hyperbola is the fit of the curve up to the deformation, where the tip starts probing the underlying PS, i.e. the fit of the part of the curve without mechanical double layer effect. All four parameters depend on x_I; in particular, the fitting parameters go from the parameters for bulk PS when $x_I = 0$ to the parameters for bulk PnBMA when $x_I > 21$ μm.

In the second hyperbola is $\alpha_2 \approx \beta_2$ and $\gamma_2 \ll \beta_2$, i.e. it is a hyperbola in which the slope of the first linear region, $\beta_2 - \alpha_2$, is nearly zero. The parameters α_2 and β_2 are constant, hence independent of x_I; γ_2 and ε_2 depend on x_I. Consequently, for the second hyperbola, the slope of the two linear regions is always the same, the slope of the first linear region being nearly zero, and only the centre and the width of the transition region, determined by ε and γ, depend on x_I; in particular, approaching the interface, the transition occurs at lower forces and is narrower.

Since the second part of the fit is subtracted from the first part, the fit in Eq. (5.1) corresponds to the fit of curves on mechanical double layers (Sect. 1.10), where the first part is additionally affected by viscoelastic behaviour (Sect. 1.8.3).

Summarising, the curves in the transition region are characterised by a "three-regime" dependence on the force:

1. In the first linear region, deformations are influenced by the underlying PS only in a very narrow stripe close to the interface. This stripe has a maximum width of about 1 μm and includes only one or two force–distance curves. At larger distances the cantilever probes only the upper PnBMA layer, since this is much thicker than the obtained deformation. The PnBMA near the interface is stiffer than far from the interface. The width of the region where the elastic modulus of PnBMA depends on x_I depends on the temperature and is some microns at 32 °C and 60 μm at 70.1 °C. Such stiffening is due to internal stresses arising from the geometrical constraints at the boundary with PS.
2. Also the beginning of the second linear region is influenced by the internal stresses.
3. At larger loads, when deformations and the ratio D/t_{PnBMA} increase, the AFM tip probes more and more the underlying PS, acting as the stiff substrate of a mechanical double layer. At very high loads, the slope of the second linear region of curves acquired in the transition region approaches the slope of the second linear region in the curves acquired on PS. With increasing distance from the interface, the second linear region becomes more and more similar to that of curves acquired on bulk PnBMA, and the interval affected by the underlying PS becomes shorter and shorter, i.e. it starts at higher and higher forces.

5.3 Hands-on Example 18: Characterisation of the Local Elastic Modulus in Confined Poly(methyl methacrylate) Films

This hands-on example reports the results of the work of Cheng et al. [2]. Aim of the experiments is to measure the mechanical properties of a confined polymer next to the interface with a stiff substrate.

Sample Preparation and Instrumentation Poly(methyl methacrylate) (PMMA) was purchased from Pressure Chemical Co. (Pittsburgh, PA) and had a molecular weight of $M_w = 387$ kDa and a polydispersity index of $M_w/M_n = 1.14$. Two different substrates were used, silica slides and alumina plates.

As illustrated in Fig. 5.7, the samples were prepared in sandwich form with silica or alumina substrate on the bottom, PMMA in the middle and a 100-μm-thick cover glass on the top. Substrates were ultrasonically cleaned in acetone for 10 min and fast dried. PMMA powder was placed between substrate and cover glass and melted at 220 °C and then pressed under a small pressure (10–50 MPa) to form the sandwich. In order to relax internal stress engendered by hot pressing, the samples were annealed after cooling at 135 °C for 24 h.

An extremely smooth surface is necessary to measure the local modulus at the nanoscale via AFM indentation. Due to the large difference of the moduli of PMMA and substrate, common mechanical polishing cannot be employed to prepare a sufficiently smooth surface over several phases. This can be achieved by means of broad-beam ion milling.

Previous to the ion-milling process, the samples were fractured in the middle and mechanically polished to obtain a relatively smooth and straight edge on which ion milling has been performed.

The instrument employed for ion milling is a Leica EM TIC3X, which is able to polish areas up to 1×4 mm^2 with three argon ion beams operating simultaneously.

Since the first 50 μm of material facing the ion beam is destroyed in the process, the cover glass on top of the sandwich samples acted as a sacrifice layer and provided protection to PMMA films. The ion-milling process was performed at a

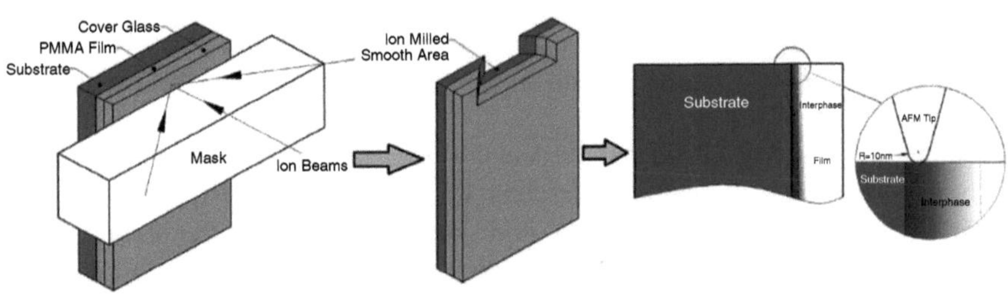

Fig. 5.7 Schematic representation of ion milling. A sandwich sample with a PMMA film between substrate and cover glass slide is milled, yielding a smooth area. The *right* part of the figure shows schematically the cross section of the final sample; measurements were performed at the PMMA-substrate interface. Adapted with permission from [2]. Copyright 2015. WILEY−VCH Verlag

temperature of −20 °C in order to avoid overheating and included two runs, the first at 5 kV for 3 h, to quickly penetrate the cover glass, and the second at 4 kV for 5 h, to finely mill the PMMA film and the substrate next to the interphase.

The measurements were performed with a Bruker Dimension ICON microscope equipped with a Bruker TAP525 cantilever with elastic constant $k_c = 200$ N/m and tip radius $R = 10$ nm. The tip radius was determined using a calibrating grid with steps and valleys of known height or depth and sharp boundaries (see Sect. 2.2.2).

Local Elastic Modulus in Confined PMMA Films Figure 5.8 shows the results of AFM indentation at the PMMA/silica interface. The top-left part of the figure shows the three-dimensional topography and the modulus map over the same 1 μm^2 area across the interface, the top-right part representative line profiles of the topography and modulus map and the bottom part the average of several modulus scan lines on three different scan areas at the PMMA/silica interface.

The topography image confirms that the surface is extremely flat with roughness less than 2 nm over 1 μm^2 area. Moreover, the silicon/PMMA boundary cannot be seen in the topography image.

The boundary between PMMA and silicon can be seen clearly in the modulus map. The interphase polymer with higher modulus is also evident as a brighter strip next to the boundary.

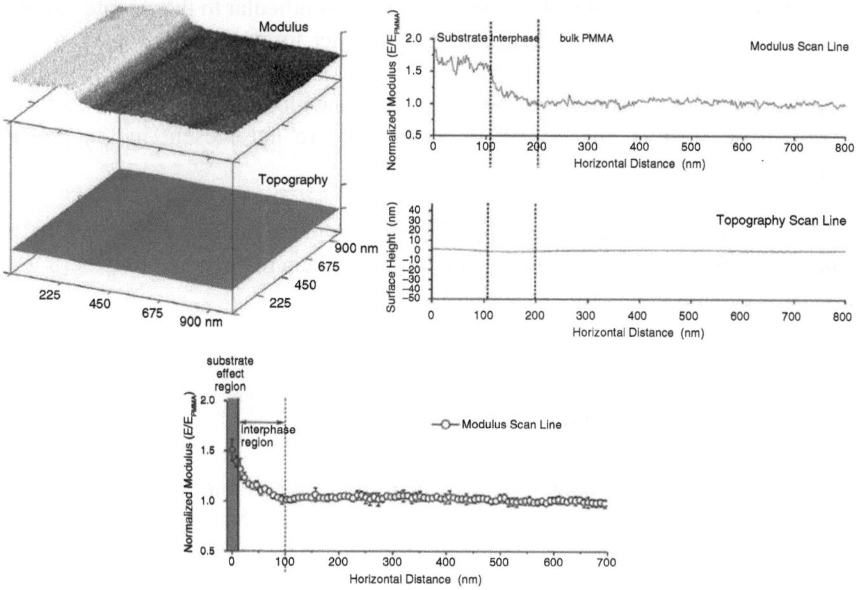

Fig. 5.8 AFM indentation results. (*Top left*) Topography and modulus maps over the same 1 μm^2 area at the PMMA/silica interface. (*Top right*) Representative scan lines of topography and modulus. (*Bottom*) Average of several modulus scan lines on three different scan areas at the PMMA/silica interface. Adapted with permission from [2]. Copyright 2015. Wiley-VCH Verlag

The modulus of bulk PMMA, obtained by fitting curves acquired far away from the boundary with DMT theory, is around 3.5 GPa. The modulus shown in Fig. 5.8 is normalised by the modulus of bulk PMMA.

In the interphase region, the value of the modulus is 1.5 times larger than that of bulk PMMA at the interface and then gradually decreases with increasing distance from the substrate. At a distance of 100 nm from the interface, the difference between the local modulus and the bulk modulus is less than 2 %.

The region in which the modulus of PMMA decreases from the value at the interface to the bulk value is wider (about 170 nm) for PMMA/alumina samples.

The higher modulus of PMMA in the interphase region can be attributed to two different phenomena. In a very narrow stripe at the interface, there is an artificial increase of the measured value of Young's modulus, either because the contact area includes both PMMA and substrate or because the stress field involves also the substrate phase. The width of this region, where the modulus is higher due to the finite size of the tip, was estimated by the authors via finite element simulations as 15 nm. At larger distances, i.e. in the actual "interphase region", the higher modulus is due again to a hard-wall effect as in the previous example. Yet, there are three fundamental differences between the present measurement and those reported in the previous example.

1. In the present experiment, the sample has been annealed after preparation. Hence, most stresses have been released before the measurements.
2. In the present measurement, the interface is perpendicular to the sample surface, and there is no substrate underneath the polymer forming a mechanical double layer.
3. In the previous measurement, the interface of the model blend was between two polymers; stresses engendered by dewetting are not present in the present experiment.

As a consequence, the stiffening of the polymer at the boundary, due only to the hard-wall effect, can be determined more precisely than in the previous experiment, even if occurring in a narrower region.

5.4 Microstructured Blends

The second part of this chapter presents measurements on microstructured blends, i.e. blends with several interfaces and whose domains have dimensions in the micron or nanometre scale.

Both microstructured blends characterised in the measurements presented in the following sections [6, 7] have been prepared through spin coating. This technique has been briefly illustrated in Sect. 4.1.

When a solution of two or more polymers is spin coated on a substrate, the structure and topography of the resulting sample are mainly determined by two processes, phase separation and dewetting.

Phase separation [8] occurs in a heterogeneous mixture, as opposed to a homogeneous mixture, when the polymers cannot mix on a molecular scale and the blend consists of several regions with different composition. Whether a mixture is homogeneous or heterogeneous is determined by Gibb's free energy of mixture ΔG_{mix}, which in turn is the sum of the mixing entropy ΔS_{mix} and mixing enthalpy ΔH_{mix}:

$$\begin{aligned}\Delta G_{\text{mix}} &= \Delta H_{\text{mix}} - T\Delta S_{\text{mix}} \\ &= R_g T\left[\chi_{12}\varphi(1-\varphi) + \frac{\varphi}{N_1}\ln\varphi + \frac{1-\varphi}{N_2}\ln(1-\varphi)\right].\end{aligned} \tag{5.2}$$

N_1 and N_2 are the polymerisation degrees of the two polymers, $\varphi = \frac{V_1}{V_1+V_2}$ and $1-\varphi = \frac{V_2}{V_1+V_2}$ are their volume fractions, R_g is the gas constant, T is the temperature and χ_{12} is Flory interaction parameter.

Flory interaction parameter is a measure of the differences between the energies of the blend before and after mixing and depends on the temperature. Usually χ_{12} is inversely proportional to temperature, corresponding to the common experience that mixing is favoured by high temperatures.

The mixing entropy is associated with motions of the polymers chains, whereas the mixing enthalpy is related to local interactions and motions of the monomers.

For polymers, N_1 and N_2 are usually large and the mixing entropy is very small. Nevertheless, it is always positive and its contribution to Gibb's free energy is always negative; hence, the entropy contribution always favours mixing.

Depending on the sign of χ_{12}, the mixing enthalpy can promote or hinder mixing. If the attraction between different monomers is stronger than the attraction between monomers of the same species, χ_{12} is negative and mixing is favoured. This case is rather an exception. Usually, the attraction between different monomers is weaker than the attraction between monomers of the same species, χ_{12} is positive and mixing is inhibited.

Depending on the ratio of the two contributions, if ΔG_{mix} is positive, the two polymers do not mix. If ΔG_{mix} is negative and the curve $\Delta G_{\text{mix}}(\varphi)$ is convex at any composition φ, the two polymers will mix, resulting in a homogeneous mixture. If ΔG_{mix} is negative, but the curve $\Delta G_{\text{mix}}(\varphi)$ is concave in a given composition range, the two polymers will undergo phase separation and form domains with two given equilibrium compositions. The composition interval in which $\Delta G_{\text{mix}}(\varphi)$ is concave is called miscibility gap and is delimited by the local minima, i.e. the compositions at which the first derivative of the free energy, $\partial\Delta G_{\text{mix}}/\partial\varphi$, is zero.

The miscibility gap encompasses unstable and metastable composition intervals, delimited by the inflection points at which the second derivative of the free energy, $\partial^2\Delta G_{\text{mix}}/\partial\varphi^2$, is zero.

In the interval between the inflection points, the system is unstable and very small fluctuations in composition are sufficient to engender the phase separation, called in this case spinodal decomposition.

Between the inflection points and the minima of the curve $\Delta G_{\mathrm{mix}}(\varphi)$, the system is metastable. Larger composition fluctuations are necessary to lead to phase separation. In this case, the separation occurs by nucleation and growth.

The second common process determining the morphology of the blend (e.g. the PS–PnBMA blend in the following hands-on example) is dewetting. Dewetting leads to the formation of holes in one of the two films.

There are two main kinds of dewetting process in thin films, heterogeneous nucleation and spinodal dewetting [9, 10]. Both mechanisms lead to the creation of holes. In the first case, the process is originated by impurities and by residual stresses in the film, and the holes are uncorrelated; in the second case, the holes result from surface ripples originated by thermal fluctuations and have a typical correlation length depending on the film-substrate interactions. Independently of the process, the holes are surrounded by rims in which the removed polymer accumulates and grow with time until they merge and disrupt the continuous film [11]. The final morphology of the sample consists of ribbons of polymer, which become unstable and build droplets.

If the film is prepared by spin coating, during evaporation of the solvent, the polymer concentration increases. As a consequence, the viscosity of the solution increases and the mobility of the polymer chains decreases. This rapid increase of the polymer concentration is analogous to a temperature quench [12]. In both cases, the reduced mobility of the chains inhibits the movements necessary for the formation of holes and ribbons [9]. As a result, the film morphology is "frozen" far from thermodynamic equilibrium.

The final morphology of a spin-coated blend depends on several factors, such as the temperature [13, 14], the blend composition [15–17] and the molecular weight of polymers [18, 19], the kind of solvent [14, 15] and of substrate [19] and the film thickness [16, 19–21].

A blend, even when it is allowed to reach the thermodynamic equilibrium, is a heterogeneous material with morphological features in the micron or nanometre range [22–24]. Knowledge of the local properties is essential for a quantitative characterisation of a polymer blend.

Several experimental techniques can be employed to study the morphology and the physical and mechanical properties of a blend [25, 26]. Yet, most of these techniques do not have the required lateral and vertical resolution in the nanometre range.

The capability of characterising the nanoscale properties of a blend with AFM force–volume measurements is shown in the following hands-on examples.

AFM measurements aimed to the characterisation of the composition and morphology of blends are performed not only in force–volume mode but also through friction force imaging and phase shift imaging [27]. Both acquiring modes have been described briefly in Sect. 1.1. Tapping Mode and in particular phase shift imaging is employed also in both following hands-on examples.

Both torsion and phase shift depend in a complex and not completely understood way on tip-sample adhesion, stiffness, viscoelasticity and sample topography [28, 29]. As a consequence, as shown in detail in the following hands-on examples, these methods provide an image contrast, but not a quantitative determination of sample properties.

5.5 Hands-on Example 19: Spatial Variation of the Properties of a Microstructured Polystyrene/Poly(*n*-butyl methacrylate) Blend

In this hands-on example, the characterisation of the mechanical properties of a microstructured blend of poly(*n*-butyl methacrylate) and polystyrene is discussed [6].

Sample Preparation and Instrumentation The granulate of poly(*n*-butyl methacrylate) (PnBMA) was purchased from Scientific Polymer Products (Ontario, NY). The molecular weight, the polydispersity index and the glass transition temperature given by the supplier are $M_w = 320$ kDa, $M_w/M_n \leq 2.58$ and $T_g = 22$ °C.

Polystyrene (PS) was purchased from BDH Chemicals (Poole, England). The molecular weight, the polydispersity index and the glass transition temperature are $M_w = 100$ kDa, $M_w/M_n \leq 3.05$ and $T_g = 98$ °C.

The blend was spin coated on glass slides, previously cleaned and rinsed with toluene. The same amount of both polymers (32.3 mg) was dissolved in 2 mL toluene. A polymer film was spin coated from 500 μL of the solution with an angular velocity $\omega = 2000$ rpm for 1 min. The samples were dried for 2 weeks under ambient conditions to obtain equilibrated and solvent-free films.

Atomic force microscopy (AFM) measurements were performed with a MFP3D microscope (Asylum Research, Santa Barbara, CA) equipped with a Pointprobe NCL cantilever (Nanosensor, Wetzlar-Blankenfeld, Germany) with elastic constant $k_c = 54 \pm 1$ N/m and a silicon tip with an estimated radius of $R = 35 \pm 5$ nm. The elastic constant was determined from the noise spectrum [3] (see Sect. 2.2.1).

Properties of a Microstructured PS–PnBMA Blend Figure 5.9 shows the topography (left) and the phase shift image (right), acquired in Tapping Mode, of a 60×60 μm^2 area of the blend. The blend consists of a non-uniform PS film with several holes on top of a uniform PnBMA film.

The morphology of the sample is the result of phase separation and dewetting.

Due to the immiscibility of the two polymers, PnBMA and PS separate, the more hydrophilic PnBMA (dark in the topography image and bright in the phase shift image) wets the glass substrate, whereas PS (bright in the topography image and dark in the phase shift image) segregates on the surface.

Furthermore, the thin PS film dewets the PnBMA at the bottom. As summarised in the introduction to this section, due to spin coating, the film morphology is "frozen" before the most holes can merge and the PS phase can build ribbons.

The bottom of the holes in the PS film is not a flat PnBMA surface; rather, small droplets of PS can be seen inside the holes. Such small domains are due again to the rapid solvent evaporation during spin coating. As the solvent evaporates, the amount of the minority component in both phases decreases, and mass transport can no longer take place because of the reduction of the polymer diffusion [14].

The thickness of the whole film has been measured with a topography image in Tapping Mode over a scratch and is 300 ± 10 nm; the thickness of the PS film is 95 ± 5 nm.

The holes in the PS film have variable dimensions and forms. The analysis of several images, with a total number of 600 holes, has shown that the mean area of the holes is 10 ± 5 μm^2. Some holes result from the merging of two holes. The result is in some cases an elongated ellipse. In some other cases, the two original holes can still be distinguished. In these two cases the circularity, defined as $p^2/(4\pi A)$, with p the perimeter and A the area, is 1.36 and 1.52, respectively. The circularity of not merged holes is 1.065, and the mean radius is 1.7 ± 0.4 μm. The mean distance of neighbour holes is 1.6 ± 0.4 μm.

The insets in both images in Fig. 5.9 show a region of the PS film, where the phase shift is larger than on the rest. The topography image reveals that this region is a light depression, 10–20 nm deeper than the residual PS film. Several depressions can be observed in both images in Fig. 5.9. They are present in each region of the PS film, where the distance between neighbour holes is larger than 2 μm. Such depressions have been observed in several simulations and experimental works [30, 31]. Their existence is probably due to thickness fluctuations [32, 33]. Also, it has been shown that, when a hole in a dewetting film has reached a certain size, "satellite" holes appear in neighbour depressions [34, 35].

Such depressions consist of PS chains, and not of PnBMA or of a mixed phase, since PnBMA and PS are immiscible and the interphase region between these two polymers, i.e. the region, where chains of both polymers coexist, has a width, depending on Flory interaction parameter χ_{12}, of some nanometres.

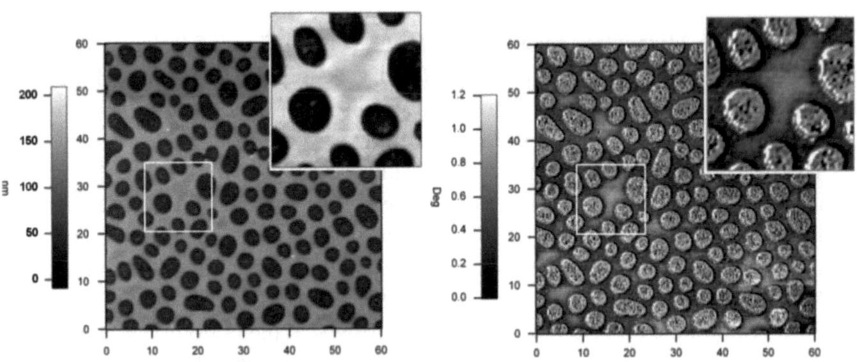

Fig. 5.9 Topography image (*left*) and phase shift image (*right*) of a 60 × 60 μm^2 area of the PS–PnBMA blend, acquired in Tapping Mode. The *white square* indicates the position of the 14 × 14 μm^2 section, magnified in the insets, where the force–volume measurement was performed. Reprinted from [6]

The phase signal in the depressions cannot be affected by the topography, since there are no steep edges and the height changes gradually. Hence, the higher phase shift in such depressions must be due to different material properties, i.e. stiffness and adhesion. Yet, the contributions of stiffness and adhesion to the phase signal cannot be distinguished. To this aim, and for a quantitative characterisation of the mechanical properties of the blend, force–distance curves are necessary.

Force–volume measurements with 100×100 curves have been performed on the $14 \times 14\ \mu m^2$ area shown in the insets of Fig. 5.9, containing a large depression. Three maps have been calculated from the force–volume: the adhesion map, the stiffness map and the map of the elastic modulus.

The map of the adhesion force F_{adh}, measured as the jump-off-contact force, is shown in the left panel of Fig. 5.10.

In this map, PnBMA regions, having a larger adhesion, are brighter than the PS film. The depression in the centre of the image has a larger adhesion than the rims around the holes, almost as large as on PnBMA.

At the border of the holes or near the PS droplets inside the holes, the adhesion assumes very large or very small values. This artefact is due to changes in the contact area or to abrupt movements of the tip (see Sect. 2.3.2).

The right panel of Fig. 5.10 shows histograms of the adhesion force. The first histogram (black line) collects all the points in the scanned area. In the other three histograms (bars), the points at the border of the holes and those on the PS droplets inside the holes, affected by artefacts, have been eliminated, and, based on the topography, the three regions of the blend have been separated: the PS rims correspond to the dark grey bars and the depression and the PnBMA film to the bright grey and white bars. The histograms can be fitted with Gaussian functions. The mean adhesion and the half width at half height are $0.42 \pm 0.08\ \mu N$ (PS rims), $0.71 \pm 0.04\ \mu N$ (PnBMA) and $0.68 \pm 0.05\ \mu N$ (depression).

From the approach contact lines of the force–distance curves, the stiffness can be calculated as $S_{eff} = k_s/(k_c + k_s)$.

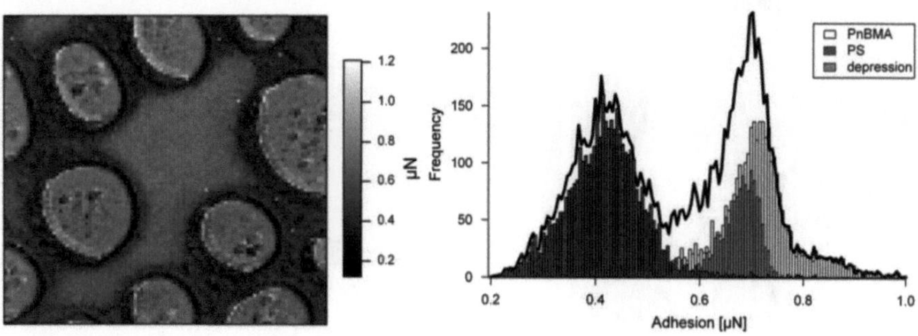

Fig. 5.10 (*Left*) Adhesion force map of the blend. The PnBMA film and the depression (*brighter*) have a larger adhesion than the PS rims around the holes. (*Right*) Histograms of the adhesion on the whole area (*black line*), on the PS rims (*dark grey bars*), on the depression (*light grey bars*) and on the PnBMA film in the holes (*white bars*). Reprinted from [6]

Figure 5.11 shows the map of the stiffness (left panel) and the corresponding histograms (right panel). As in Fig. 5.10, the black line is the histogram of all points, whereas the dark grey, the bright grey and the white bars are the histograms of the points on the PS film, on the depression and on the PnBMA film, respectively.

In this map, the PnBMA regions, which are more compliant, are darker than the PS rims around the holes. The stiffness of the depression is intermediate between that of PnBMA and the rest of the PS film. The borders of the depression are not well defined as in the adhesion and in the phase shift map, since some parts of the depression, most of all the bottom, narrow region, which can be clearly identified as a portion of the depression in Fig. 5.10, have nearly the same stiffness as the rest of the PS film. The histograms of the stiffness are shown in the right panel. The fit with Gaussian functions yields a mean stiffness $S_{\mathrm{eff}} = 0.89 \pm 0.03$ for PS in the rims, $S_{\mathrm{eff}} = 0.81 \pm 0.04$ for PnBMA and $S_{\mathrm{eff}} = 0.85 \pm 0.02$ for the depression.

Figure 5.12 shows the deformation–load curves obtained by averaging the curves in the three regions of the blend. The three curves can be fitted very well with Hertz equation for a hemispherical tip, $D = \left(\frac{F}{E_{\mathrm{tot}}\sqrt{R}}\right)^{2/3}$ (Eq. 1.19). From the shape of the curves, it is evident that the glass substrate does not affect the mechanical properties of the PnBMA film, due to its thickness (compare Sect. 4.2). A double-layer effect can be excluded also for the PS film, both in the rims around the holes and in the depression, since the thickness of the film in both region is much larger than its maximum deformation (95 and 85 nm compared to 9 and 13 nm).

The mean values of the elastic moduli out of nine measurements on different samples (prepared in the same way) with different cantilevers are 3.1 ± 0.3 GPa for PnBMA, 10.6 ± 0.1 GPa for PS and 5.6 ± 0.4 GPa for the depression.

The elastic moduli of PnBMA and PS measured in this experiment are larger than the literature values of ca. 1 and 3 GPa, but the value for PnBMA is in agreement with the measured values of spin-coated films of PnBMA (see

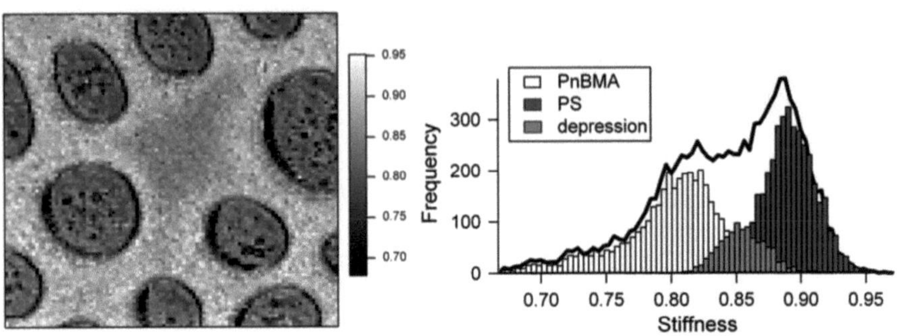

Fig. 5.11 (*Left*) Stiffness map of the blend section. PnBMA (*darker*) is more compliant than PS in the rims around the holes. The depression shows an intermediate stiffness. (*Right*) Histograms of the stiffness on the whole area (*black line*), on the PS rims around the holes (*dark grey bars*), on the depression (*light grey bars*) and on the PnBMA film in the holes (*white bars*). Reprinted from [6]

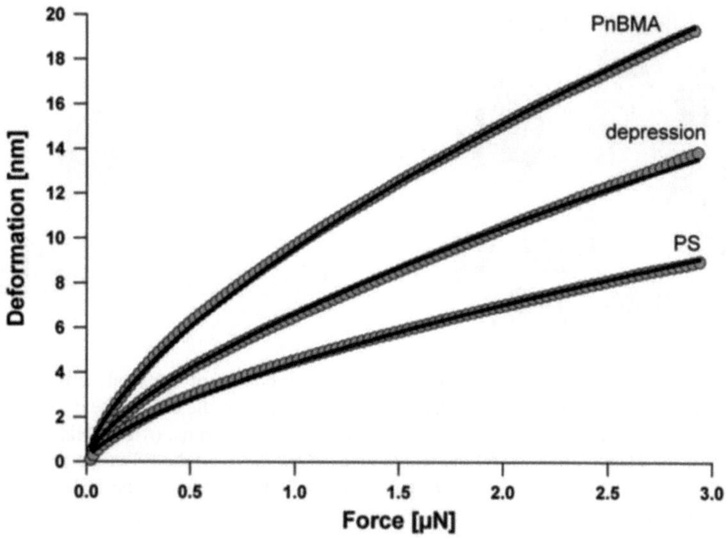

Fig. 5.12 Averaged deformation–force curves (*grey circles*) on PnBMA (*upper curve*), the depression (*middle curve*) and PS (*bottom curve*). The curves are fitted with Hertz equation (*black line*). Reprinted from [6]

Table 3.1 in Sect. 3.1). The larger elastic moduli are due to the spin-coating procedure (compare experiments reported in Sects. 4.2 and 4.3) and to the constraints of the polymers in the blend. The fact that PnBMA has the same modulus as spin-coated films proves that there are no additional internal stresses due to the vicinity of PS. For the stiffening of PS, other factors cannot be excluded. The dewetting mechanisms probably engender inner stresses leading to an increase of the elastic modulus. Since PS at ambient temperature is well under its glass transition temperature, the chains cannot relax the inner stresses caused by the sample preparation and the contact with PnBMA.

The elastic modulus of the blend section has been calculated for each force–distance curve. The resulting map and histograms are shown in Fig. 5.13.

As for the adhesion and the stiffness, the three histograms can be separated and the depression has an intermediate value of the elastic modulus.

It is clear that the adhesion correlates with the phase shift, but not with the stiffness and the elastic modulus; in particular, the adhesion and the phase shift of the depression are indistinguishable from those of PnBMA, whereas the elastic modulus is intermediate between those of PnBMA and PS. Furthermore, the depression is much smaller in the map of the elastic modulus than in those of the adhesion and of the phase shift, since an approximately 1-µm-wide portion at the border has the same modulus as PS in the rims, but not the same adhesion and phase shift.

This observation has some important consequences. First of all, it can be assessed that the dissipated energy and the phase shift in the present experiment are dominated by the adhesion and not by the mechanical properties of the sample.

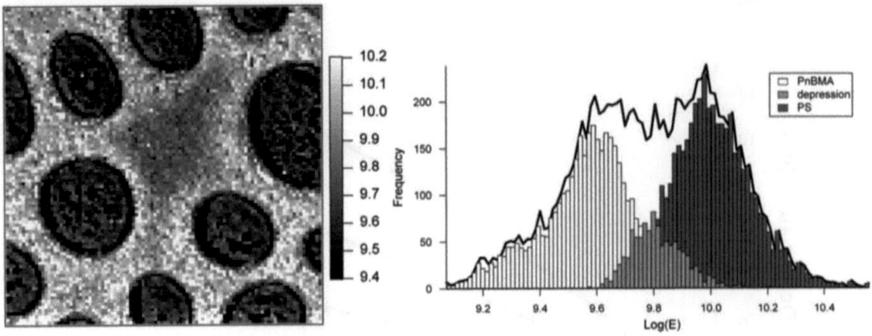

Fig. 5.13 (*Left*) Map of the logarithm of the elastic modulus of the analysed blend section. PnBMA (*darker*) has a lower modulus than PS in the rims around the holes. The depression has an intermediate modulus between that of PnBMA and of PS in the rims. (*Right*) Histograms of the logarithm of the elastic modulus on the whole area (*black line*), on the PS in the rims (*dark grey bars*), on the depression (*light grey bars*) and on the PnBMA film in the holes (*white bars*). Reprinted from [6]

The tip-sample adhesion depends primarily on the tip-sample interactions, but also on the contact area and on the mobility of the sample chains. The interactions between the tip and PS are the same in the rims and in the depression, and they cannot engender any difference in the adhesion. The contact area, being proportional to the deformation, depends on the elastic modulus and is different for the three regions of the sample. Nevertheless, it affects both the stiffness and the adhesion and cannot be the cause for the different dimensions of the depression in the stiffness and adhesion maps.

Yet, it is known that depressions are precursors of new holes, which would have formed, if the condition of the film had not been frozen by the abrupt solvent evaporation. It can also be assumed that the mobility of the PS chains in the depressions is higher than that of the chains in the rims.

The stiffness and the elastic modulus are measured along the approach contact line, and in this case the PS chains are pushed to contact with the underlying PnBMA film, whereas the adhesion is measured in the withdrawal curve, when the tip is retracted from the sample and the PS chains adhering to the tip are pulled away from the PnBMA substrate.

This permits to assume that the higher adhesion, the lower stiffness and the lower elastic modulus of PS in the depression compared to PS in the rims are due to the higher mobility of the chains in the depression. Furthermore, the effect of the chain mobility on the adhesion is higher than on the stiffness and on the elastic modulus because, when measuring the stiffness, the PS chains are compelled to contact the PnBMA substrate, whereas, when measuring the adhesion, they detach from the PnBMA substrate and follow the tip. This is the reason for the larger dimensions of the depression in the adhesion map, compared to the map of Young's modulus.

5.6 Hands-on Example 20: Spatial Variation of the Properties of a Polystyrene/Polybutadiene Blend

This hands-on example deals with a part of the measurements exposed in [7]. The characterisation of the mechanical properties of the homogeneous polybutadiene (PB) film is outlined in Sect. 3.3; measurements concerning the mechanical polystyrene/polybutadiene double layer and the thickness of the PB layer on top of the polystyrene (PS) domains are the subject of Sect. 4.5. Sample preparation and instrumentation have been illustrated in Sect. 4.5.

Adhesion, Dissipated Energy and Stiffness During spin coating onto the glass substrate, PS and PB, which are immiscible, undergo phase separation. The result is a microstructured blend with PS domains in a PB matrix. Yet, PB wets the PS phase. Experiments concerning the mechanical properties of the PB–PS double layer and the thickness of the PB layer on top of the PS domains are discussed in Sect. 4.5. In this example, measurements of the adhesion, dissipated energy and stiffness of the sample are reported.

The film thickness of the blend has been measured by acquiring a topography image in Tapping Mode over a scratch and resulted to be 275 ± 20 nm, with the domains emerging for a maximum of further 150 nm from the matrix.

Figure 5.14 shows a 20×20 μm^2 topography image (A) acquired in Tapping Mode and the corresponding phase shift image (B). The cantilever used in this measurement had a spring constant $k_c = 50$ N/m.

The blend consists of a PB matrix and PS domains of different sizes emerging from the matrix [36]. The diameter of the domains varies between some 10 nm and some 10 μm. The small domains are round, whereas larger domains have different shapes resulting from the mergence of two or more round domains. On the large domains, there are some small holes with diameters between 50 nm and 1 μm.

The difference in phase shift between PB and PS is much larger than the deviations inside the homogeneous regions; hence, the phase shift image shows a black and white contrast between the matrix and the small holes on one side and the domains on the other side.

As already seen in the previous example, Tapping Mode yields only qualitative information about the sample properties and does not permit to distinguish between the adhesion and the mechanical contributions to the phase shift. Hence, force–distance curves were used for a quantitative study of mechanical properties, adhesion and dissipated energy.

Figure 5.15 shows the stiffness (A) and hysteresis (B) maps calculated from a force–volume with 100×100 curves. The employed cantilever had an elastic constant $k_c = 8$ N/m and a tip radius $R = 40$ nm. In both maps there is a distinct material contrast between the two polymers.

Since PS is stiffer than PB, in the stiffness map, showing the quantity $S_{\text{eff}} = k_s/(k_c + k_s)$, the PS domains appear brighter, i.e. stiffer, than the PB matrix, which is almost uniformly black. Some regions, assuming different grey values, have stiffness values between 0.1 (PB matrix) and 0.7 (PS domains). One of these

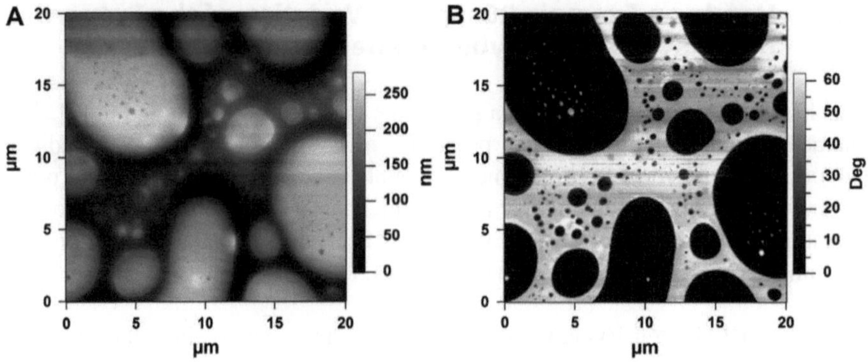

Fig. 5.14 Tapping Mode topography (**A**) and phase shift image (**B**) of the PB–PB blend. PS forms emerging domains surrounded by a PB matrix. Reprinted with permission from [7]. Copyright 2013. Elsevier

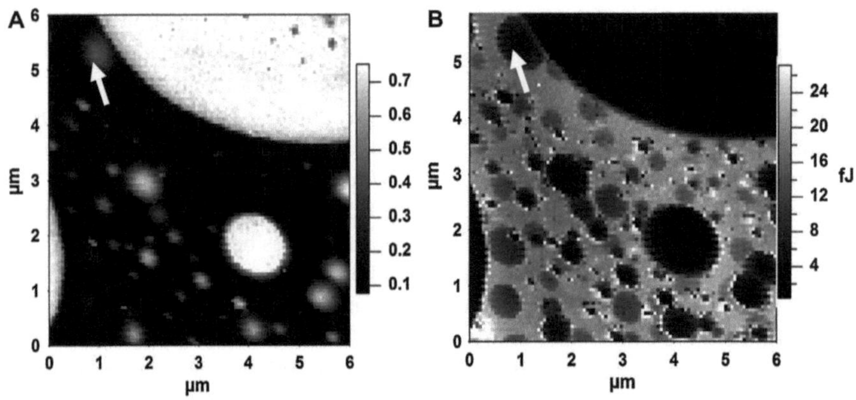

Fig. 5.15 Stiffness (**A**) and hysteresis (**B**) maps of the blend. In the stiffness map, *black* regions represent the PB matrix, *white* regions the PS domains and *grey* regions hidden PS domains. In the hysteresis map, *light grey* regions correspond to the PB matrix, *black* regions to the PS domains and *intermediate grey* regions to hidden PS domains. Reprinted with permission from [7]. Copyright 2013. Elsevier

regions is highlighted with an arrow. Such regions are low and small PS domains, which do not emerge from the PB matrix (see Sect. 4.5). As a consequence, they cannot be seen in the phase shift image because in Tapping Mode the tip does not indent the sample.

The hysteresis is calculated as the area between approach and retraction curve above the axis $\delta = F = 0$. Since the area below the axis $F = 0$ is not included in the calculation, this hysteresis does not take into account the adhesion. Rather, this area is proportional to the energy dissipated in a loading–unloading cycle due to deformation of the sample and is larger than zero in presence of plastic deformations, as defined in Sect. 1.9. In the hysteresis map, the dissipated energy on PB (ca. 20 fJ) is larger than on PS (ca. 5 fJ). The adhesion map, i.e. the map collecting the values of the jump-off-contact, shows only small differences between PS and PB. It can be concluded that differences in phase shift are due mostly to the larger deformation of PB.

In the hysteresis map, hidden domains can be distinguished better and are more numerous than in the stiffness map. This is due to the fact that the stiffness is calculated via a linear fit, whereas the hysteresis is the result of an integration and is less affected by noise.

From the measurements described in Sect. 4.5, it is known that the PS domains are covered with a thin layer of PB. The thin PB layer on top of the PS domains explains the absence of a contrast in the adhesion maps, since the interaction determining the jump-off-contact and the adhesion is everywhere on the blend the interaction with PB. The presence of a thin PB layer on the whole sample is also the reason why it is not possible to calculate the elastic moduli of both materials in a measurement with a unique cantilever. The elastic modulus of PB can be measured only with a compliant cantilever, but, for such a cantilever, the PS domains are mechanical double layers, as shown in Sect. 4.5, and the determination of the elastic modulus of PS is affected by large uncertainties. On the other hand, when using a very stiff cantilever, the PB matrix is pierced by the tip and uncertainties result from very large deformations.

Improving the Lateral Resolution of Force–Volume Measurements In [7], further measurements are aimed to the optimization of the lateral resolution in force–volume measurements.

The lateral resolution is defined as the distance between two successive force–distance curves. The maximum resolution achieved in this work was 6 nm and to date the best resolution achieved within a force–volume measurement [37–42].

Such a resolution has been reached with a cantilever with spring constant $k_c = 24$ N/m, in a force–volume measurement with 100×100 curves on a 600×550 nm^2 area.

A fundamental condition for a high resolution in a force–volume measurement is that force–distance curves do not influence each other, i.e. that a curve does not indent the region deformed during the acquisition of the previous curve. To this aim, both tip radius and indentation depth must be very small. In this measurement, the tip radius was smaller than 15 nm and the maximum indentation depth was between 18 nm on the PS domain and 200 nm on the PB matrix.

The large indentation depth on PB had no effect on the resolution of the measurement, since PB, which is well above its glass transition temperature at room temperature, is able to regain completely its form even in the short time elapsing between two indentations.

The indentation depth on the PS domains (18 nm) is quite large when compared with the distance between two successive curves (6 nm). Yet, in the

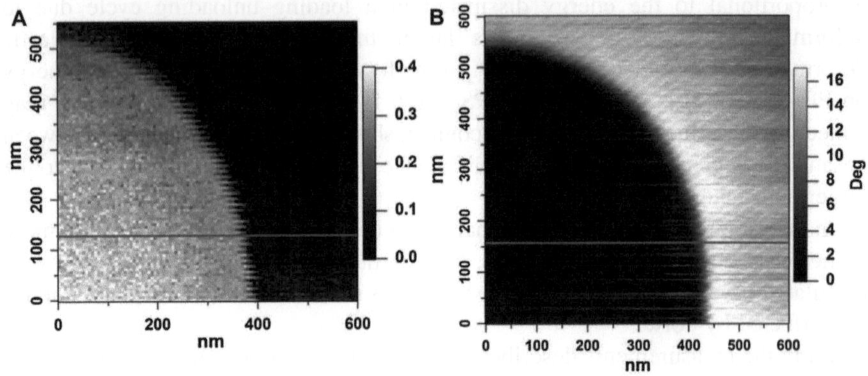

Fig. 5.16 Map of the stiffness of the blend with a resolution of 6 nm (**A**) and a phase shift image acquired on the same area shortly before (**B**). Differences in the dimensions of the two images are due to drift during the force–volume. Reprinted with permission from [7]. Copyright 2013. Elsevier

first 15 nm, the tip is piercing through the PB layer on top of the domains; only in the last 3 nm the tip indents the PS domain. As a consequence, PS was not deformed plastically.

Figure 5.16 shows the map of the stiffness S_{eff} calculated from this force–volume (A) and a phase shift image acquired on the same area shortly before (B).

In this measurement, Young's modulus of PB was 4.0 ± 0.1 MPa, which is consistent with the value of unoxidised PB reported in Sect. 3.3. As explained before, due to the PB–PS double layer and the short indentation, it is impossible to calculate the elastic modulus of PS. Force–distance curves with larger indentation have been acquired with the same cantilever on the same sample, yielding $E = 7.4 \pm 0.6$ GPa for PS. Also this value is consistent with previous measurements (compare Sect. 4.5).

The alternating lines visible at the interface in the stiffness map are the result of small amounts of PB sticking onto the tip during the acquisition of the first two or three force–distance curves on the PS domains. Such a thin PB layer affects the stiffness, as the tip has to pierce it before indenting PS. This artefact is not present in the lines where the tip is moving from PS to PB.

In this force–volume curves have been acquired with a speed of 3.8 μm/s and a frequency of 1.3 Hz, and the whole measurement took about 2 h. In such a long time, the thermal drift leads to a distortion of the slow scanning dimension, i.e. the Y-axis. The distortion is linear (0.5 nm per scan line) and can be corrected by multiplying the Y-axis by 5.5/6. Hence, the thermal drift does not affect the resolution of the measurement.

Artefacts due to the thermal drift could be eliminated through a reduction of the acquisition time, i.e. through increase of the acquisition frequency of the curves. Yet, also large acquisition frequencies (>4 Hz) cause several artefacts [43, 44].

To compare the resolution of the force–volume with that of Tapping Mode, profiles of the stiffness S_{eff} are compared in Fig. 5.17 with profiles of topography

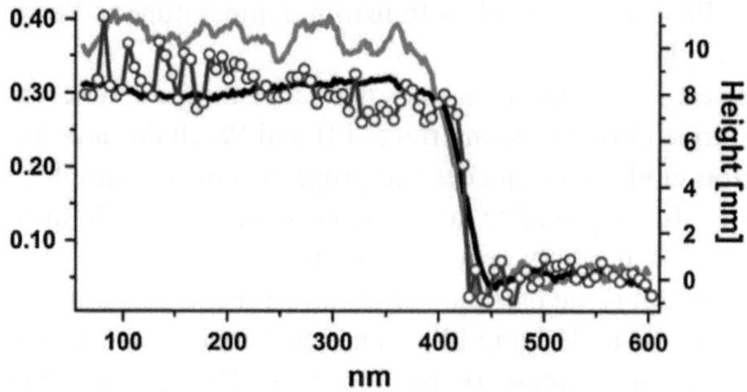

Fig. 5.17 Profiles of phase shift (*black*), topography (*grey*, referred to the right *Y*-axis) and stiffness (*grey* with markers, referred to the left *Y*-axis). All profiles are along the *grey lines* in Fig. 5.16. The phase shift profile has been normalised and is represented in arbitrary units. Reprinted with permission from [7]. Copyright 2013. Elsevier

and phase shift obtained in Tapping Mode. The grey line with markers represents the stiffness, the grey line the topography and the black line the phase shift. The phase shift profile, calculated out of three lines in the Tapping Mode image, has been normalised and is represented in arbitrary units. The location of the profiles is shown in Fig. 5.16 with a grey line.

The transition region from PB to PS in all three profiles is less than 20 nm wide (even slightly narrower in the stiffness profile). Hence, the resolution of the force–volume is at least as good as in Tapping Mode.

There are three possible reasons for the width of the transition region: (1) the width results from the convolution between the tip shape and the sample topography, (2) the real width is smaller than the contact area or (3) the measured width of the transition region corresponds to the real width.

The first artefact, described in Sect. 2.3.2, can be excluded, since convolution affects the measurement only when the slope of the sample topography is larger than the slope of the tip. This is not the case for the domain in Fig. 5.16, which is just 10 nm higher than PB and has no steep borders.

The second phenomenon has been discussed in detail in Sect. 2.3.1. If the interface between PS and PB is narrower than 20 nm, the width of the transition region in the measured profiles is due to the fact that, at the interface, both polymers are present in the contact area. For example, in this case, the stiffness would decrease near the interface from left to right due to the increased portion of PB underneath the tip. Yet, the widths of the transition regions in the profiles of phase shift, stiffness and hysteresis are nearly the same. If the width was affected by the contact area, it should be concluded that the contact areas in Tapping Mode and force–volume are the same. This is improbable, as in Tapping Mode the tip indents the sample less deep than in a force–volume measurement.

Hence, the profiles represent most likely the real sample properties.

This third case can be explained by several phenomena. First, PB near the interface could be stiffer than PB far away from it, due to residual stresses.

However, since PB is above its glass transition temperature and can relax stresses, this explanation is improbable.

A second possibility is the presence of a large mixture zone or interphase. In such a narrow stripe close to the interface, PB and PS chains are intermingled, and the concentration gradient influences the properties of the sample, e.g. the elastic modulus. This is also improbable, as the interphase between immiscible polymers (like PS and PB) is commonly only some nanometres wide.

A third explanation is the presence of PS underneath the PB matrix. In this case, the tip would sense the underlying PS when indenting PB close to the interface, and this would lead to an increase of the stiffness. This phenomenon is the most plausible, since the domains underneath the PB matrix have a round shape (see Sect. 4.5).

Several factors make a substantial improvement of the resolution beyond 6 nm unrealistic.

As already said, the major problem is to ensure that adjacent force–distance curves do not influence each other. To this aim, either the sample must be deformed only elastically or the width of the plastically deformed zone must be reduced by decreasing the tip size.

In order to indent the sample only elastically, either the indentation depth must be very small or the tip must be rather large, compared to the indentation depth. Yet, the indentation depth cannot be reduced indefinitely. For example, if a resolution of 2 nm is aimed, the deformation should be some angstroms. Such a small deformation would lead to a very low signal-to-noise ratio. Furthermore, the sample could be no longer assumed as a continuum, and elastic continuum theories could not be used to calculate Young's moduli.

The use of a larger and less sharp tip allows higher indentation depth. Yet, a larger tip is accompanied by a loss in resolution, since also the contact radius increases and the contact areas might overlap.

Reducing the tip radius down to some nanometres also presents some problems. The characterisation of the geometry and dimensions of a very small tip is very difficult. Moreover, a very small tip is very fragile and could be damaged during the measurement. Also in this case the indentation depth should be very small and the assumption that the sample behaves as a continuum would be no longer valid.

5.7 Pulsed Force Mode

An evident drawback of force–volume mode is the long time required for the acquisition and analysis of data. This is not only a practical disadvantage, since, as seen in the last hands-on example, during such long acquisition times, the sample thermally drifts, leading to distortions of the scanned area. Moreover, collecting hundreds or thousands of force–distance curves with thousands of points for each curve means saving very large data files, which may be an additional problem for the analysis.

One of the most effective alternative scanning modes aimed to overcome the long acquisition and analysis times is the pulsed force mode (PFM), first implemented in the group of Marti [45, 46]. PFM can be added as an external module to any AFM, provided that the feedback signal from the photo detector is accessible.

In this mode, while the *XY*-scanning is performed as usual, the *Z*-piezo elongation is modulated through a sinusoidal voltage with a frequency well below the resonance frequency of the cantilever and amplitudes ensuring that the cantilever is in and out of contact during each oscillation cycle (typically 10–500 nm).

Figure 5.18 shows a schematic representation of the modulation voltage (dotted line) and the force signal (continuous line) over a cycle.

During a cycle, the cantilever goes through a complete force–distance interval from non-contact to the maximum force F_{max}. Yet, since in force–volume mode the common frequency is between 1 and 4 Hz and in PFM is between 0.1 and 5 kHz, the measurement is at least 100 times faster. Furthermore, to bypass problems due to the volume of the saved data files and to make the analysis faster and easier, only the following features of the force signal are extracted:

1. The value of the force in non-contact regime, corresponding to the zero line of a conventional force–distance curve. Oscillations of the cantilever after the detachment may affect the measurement of this value.
2. The maximum force F_{max}; the value of the piezo elongation at F_{max} yields the sample topography.
3. The minimum value of the force signal, representative of the maximum adhesion force.

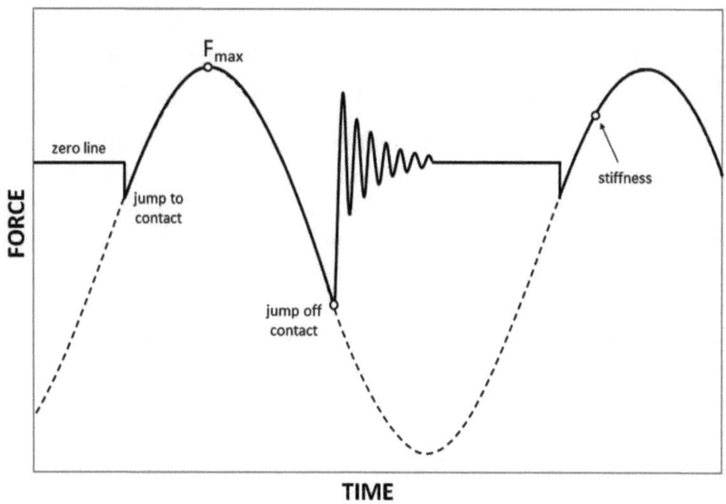

Fig. 5.18 Schematic representation of the modulation voltage (*dotted line*) and the force signal (*continuous line*) over a period in pulsed force mode. *Circles* indicate the points at which adhesion, stiffness and F_{max} are measured

4. The force difference between a point in the repulsive part of the force signal, corresponding to the contact line of a conventional force–distance curve, and F_{max}, yielding an estimation of the sample stiffness.

These four features of the force–distance cycle can be mapped to get images, which allow characterising the morphology of heterogeneous samples.

Yet, it is important to point out that in PFM, even if the acquisition and the analysis is speeded up, since a whole force–distance curve is reduced to four points, several important details are lost: information about deformation and elastic–plastic behaviour of the sample, about additional minima in the retraction curve, about jump-off-contact or slide-off-contact, about the distance dependence of forces prior to contact, etc. Hence, quantitative measurements of sample properties are often not possible; in particular, the determination of the elastic modulus of the sample requires a fit of the deformation–force curve, which can be performed only by acquiring the whole contact line [47].

Such limitation has been overcome by further developments of PFM, in which the whole deflection or force signal is recorded with high time resolution. This recently improved technique is usually called PeakForce™ mode.

PFM has been used for the characterisation of different homogeneous and heterogeneous polymer samples [46–49]. In particular, in [48], the temperature dependence of adhesion, but not of the elastic modulus, of polystyrene samples with different molecular weight (see Sect. 3.13) has been measured also in PFM mode. In [46], one of the studied samples is a microstructured polystyrene/poly(methyl methacrylate) blend; results obtained on this sample with PFM are similar to those discussed in Sect. 5.5.

With PeakForce™ mode, Young et al. [50] have measured the elastic moduli of 12 different polymers, acquiring force–distance curves with three different cantilevers. Curves with a very small indentation (2 nm) are analysed with DMT theory, i.e. the reduced elastic modulus is calculated from the equation $E_{tot} = \frac{3(F_{max} - F_{adh})}{4\sqrt{RD^3_{max}}}$ from the measured values of the maximum force F_{max}, of the adhesion force F_{adh} and of the maximum indentation D_{max}. Results are compared with nanoindenter measurements and values provided by manufacturers. The authors show that the PeakForce™ mode is able to provide repeatable measurements of the moduli of different polymers. Yet, there are mainly three limitations:

1. The indentation is always very small (some nanometres), so that measurements are strongly affected by the sample topography (see Sect. 2.3.2).
2. Quantitative results are based on calibration procedures performed on a certain material (in this case polystyrene), and difficulties arise when the studied polymers have moduli differing significantly from those of the calibration materials.
3. The analysis software cannot be modified and curves must always be analysed with DMT theory, even when JKR theory would be more suitable.

References

1. Cappella B, Kaliappan SK (2006) Determination of thermomechanical properties of a model polymer blend. Macromolecules 39:9243–9252
2. Cheng X, Putz KW, Wood CD, Brinson LC (2015) Characterization of local elastic modulus in confined polymer films via AFM indentation. Macromol Rapid Commun 36:391–397
3. Hutter JL, Bechhoefer J (1993) Calibration of atomic-force microscope tips. Rev Sci Instrum 64:1868–1878
4. Cappella B, Kaliappan SK, Sturm H (2005) Using AFM force–distance curves to study the glass-to-rubber transition of amorphous polymers and their elastic–plastic properties as a function of temperature. Macromolecules 38:1874–1881
5. Kaliappan SK, Cappella B (2005) Temperature dependent elastic–plastic behaviour of polystyrene studied using AFM force–distance curves. Polymer 46:11416–11423
6. Cappella B (2011) Mechanical properties and adhesion of a micro structured polymer blend. Polymers 3:1091–1106
7. Krämer G, Griepentrog M, Bonaccurso E, Cappella B (2014) Study of morphology and mechanical properties of polystyrene–polybutadiene blends with nanometer resolution using AFM and force–distance curves. Eur Polym J 55:123–134
8. Rubinstein M, Colby RH (2003) Polymer physics. Oxford University Press, Oxford
9. Bäumchen O, Jacobs K (2010) Slip effects in polymer thin films. J Phys Condens Matter 22:1–21
10. Seemann R, Herminghaus S, Neto C, Schlagowski S, Podzimek D, Konrad R, Mantz H, Jacobs K (2005) Dynamics and structure formation in thin polymer melt films. J Phys Condens Matter 17:S267–S290
11. Volegova IA, Buzin AI (2007) Morphology of ultrathin polymer films based on poly (ethylene oxide) blends. Polym Sci A 49:1014–1019
12. Dalnoki-Veress K, Forrest JA, Stevens JR, Dutcher JR (1997) Phase separation morphology of spin-coated polymer blend thin films. Phys A 239:87–94
13. Ton-That C, Shard AG, Daley R, Bradley RH (2000) Effects of annealing on the surface composition and morphology of PS/PMMA blend. Macromolecules 33:8453–8459
14. Walheim S, Böltau M, Mlynek J, Krausch G, Steiner U (1997) Structure formation via polymer demixing in spin-cast films. Macromolecules 30:4995–5003
15. Cui L, Ding Y, Li X, Wang Z, Han Y (2006) Solvent and polymer concentration effects on the surface morphology evolution of immiscible polystyrene/poly (methyl methacrylate) blends. Thin Solid Films 515:2038–2048
16. Ton-That C, Shard AG, Bradley RH (2002) Surface feature size of spin cast PS/PMMA blends. Polymer 43:4973–4977
17. Müller-Buschbaum P, Gutmann JS, Stamm M (2000) Influence of blend composition on phase separation and dewetting of thin polymer blend films. Macromolecules 33:4886–4895
18. Geoghegan M, Jones RAL, Clough AS, Penfold J (1995) The morphology of as-cast films of a polymer blend—dependence on polymer molecular-weight. J Polym Sci B 33:1307–1311
19. Tanaka K, Takahara A, Kajiyama T (1998) Surface molecular aggregation structure and surface molecular motions of high-molecular-weight polystyrene low-molecular-weight poly (methyl methacrylate) blend films. Macromolecules 31:863–869
20. Tanaka K, Takahara A, Kajiyama T (1996) Film thickness dependence of the surface structure of immiscible polystyrene/poly (methyl methacrylate) blends. Macromolecules 29:3232–3239
21. Wang H, Composto RJ (2003) Wetting and phase separation in polymer blend films: identification of four thickness regimes with distinct morphological pathways. Interf Sci 11:237–248
22. Paul DR, Newman S (eds) (1978) Polymer blends. Academic, New York
23. Paul DR, Barlow JW (1980) Polymer blends (or alloys). J Macromol Sci Rev Macromol Chem Phys C18:109–168
24. Paul DR, Bucknall CB (eds) (2000) Polymer blend. Wiley, New York

25. Affrossman S, Bertrand P, Hartshorne M, Kiff T, Leonard D, Pethrick RA, Richards RW (1996) Surface segregation in blends of polystyrene and perfluorohexane double end capped polystyrene studied by static SIMS, ISS, and XPS. Macromolecules 29:5432–5437
26. VanLandingham MR, Villarrubia JS, Guthrie WF, Meyers GF (2001) Nanoindentation of polymers: an overview. Macromol Symp 167:15–43
27. Paige MF (2003) A comparison of atomic force microscope friction and phase imaging for the characterization of an immiscible polystyrene/poly(methyl methacrylate) blend film. Polymer 44:6345–6352
28. Magonov SN, Elings V, Whangbo MH (1997) Phase imaging and stiffness in tapping-mode atomic force microscopy. Surf Sci 375:L385–L391
29. Noy A, Sanders CH, Vezenov DV, Wong SS, Lieber CM (1998) Chemically-sensitive imaging in tapping mode by chemical force microscopy: relationship between phase lag and adhesion. Langmuir 14:1508–1511
30. Sharma A (2003) Many paths to dewetting of thin films: anatomy and physiology of surface instability. Eur Phys J E 12:397–407
31. Neto C, Jacobs K, Seemann R, Blossey R, Becker J, Grün G (2003) Satellite hole formation during dewetting: experiment and simulation. J Phys Condens Matter 15:3355–3366
32. Seemann R, Herminghaus S, Jacobs K (2001) Shape of a liquid front upon dewetting. Phys Rev Lett 87:196101
33. Herminghaus S, Seemann R, Jacobs K (2002) Generic morphologies of viscoelastic dewetting fronts. Phys Rev Lett 89:56101
34. Becker J, Grün G, Seemann R, Mantz H, Jacobs K, Mecke KR, Blossey R (2003) Complex dewetting scenarios captured by thin-film models. Nat Mater 2:59–63
35. Chen XY, Lin J, Liu JM, Liu ZG (2009) Formation and evolution of self-organized hexagonal patterns on silicon surface by laser irradiation in water. Appl Phys A 94:649–656
36. Raghavan D, Gu X, Nguyen T, VanLandingham M, Karim A (2000) Mapping polymer heterogeneity using atomic force microscopy phase imaging and nanoscale indentation. Macromolecules 33:2573–83
37. An H, Nussio MR, Huson MG, Voelcker NH, Shapter JG (2010) Material properties of lipid microdomains: force–volume imaging study of the effect of cholesterol on lipid microdomain rigidity. Biophys J 99:834–844
38. Arce FT, Avci R, Beech IB, Cooksey KE, Wigglesworth-Cooksey B (2003) Microelastic properties of minimally adhesive surfaces: a comparative study of RTV11 (TM) and intersleek elastomers (TM). J Chem Phys 119:1671–1682
39. Gao SL, Mader E, Zhandarov SF (2004) Carbon fibers and composites with epoxy resins: topography, fractography and interphases. Carbon 42:515–529
40. Nitta T, Haga H, Kawabata K, Abe K, Sambongi T (2000) Comparing microscopic with macroscopic elastic properties of polymer gel. Ultramicroscopy 82:223–226
41. Reynaud C, Sommer F, Quet C, El Bounia N, Duc TM (2000) Quantitative determination of Young's modulus on a biphase polymer system using atomic force microscopy. Surf Interface Anal 30:185–189
42. Wang D, Fujinami S, Liu H, Nakajima K, Nishi T (2010) Investigation of reactive polymer-polymer interface using nanomechanical mapping. Macromolecules 43:5521–5523
43. Butt H-J, Cappella B, Kappl M (2005) Force measurements with the atomic force microscope: technique, interpretation and applications. Surf Sci Rep 59:1–152
44. Cappella B, Dietler G (1999) Force-distance curves by atomic force microscopy. Surf Sci Rep 34:1–104
45. Rosa-Zeiser A, Weilandt E, Hild S, Marti O (1997) The simultaneous measurement of elastic, electrostatic and adhesive properties by scanning force microscopy: pulsed-force mode operation. Meas Sci Technol 8:1333–1338
46. Krotil HU, Stifter T, Waschipky H, Weishaupt K, Hild S, Marti O (1999) Pulsed force mode: a new method for the investigation of surface properties. Surf Interf Anal 27:336–340

47. Rezende CA, Lee LT, Galembeck F (2009) Surface mechanical properties of thin polymer films investigated by AFM in pulsed force mode. Langmuir 25:9938–9946
48. Marti O, Stifter T, Waschiphy H, Quintus M, Hild S (1999) Scanning probe microscopy of heterogeneous polymers. Colloids Surf A 154:65–73
49. Grandy DB, Hourston DJ, Price DM, Reading M, Silva GG, Song M, Sykes PA (2000) Microthermal characterization of segmented polyurethane elastomers and a polystyrene-poly (methyl methacrylate) polymer blend using variable temperature pulsed force mode atomic force microscopy. Macromolecules 33:9348–9359
50. Young TJ, Monclus MA, Burnett TL, Broughton WR, Ogin SL, Smith PA (2011) The use of the PeakForce™ quantitative nanomechanical mapping AFM-based method for high-resolution Young's modulus measurement of polymers. Meas Sci Technol 22:125703

Creep Compliance Measurement

6

Abstract

There are to date very few experiments reporting the measurement of creep compliance or of creep curves with AFM, since most measurements in this field are performed with instrumented nanoindenters and experimental protocols for the acquisition of creep curves are usually not implemented in commercial microscopes. Yet, AFM offers two significant advantages: a better speed performance and a lower thermal drift. For example, the stepping time of an AFM can be smaller than 1 ms, whereas the stepping time of nanoindenters is commonly limited to ca. 1000 ms (Braunsmann et al., Polymer 55:219–225, 2014).

This section presents two hands-on examples. In the first one (Moeller, J Pol Sci B Pol Phys 47:1573–1587, 2009), creep measurements are compared with force–distance curves measurements analysed with Oliver and Pharr method. Limitations of both methods, mainly due to the occurrence of plastic deformations, are surveyed.

In the second hands-on example (Braunsmann et al., Polymer 55:219–225, 2014), the unique feature of AFM, i.e. the possibility of scanning the sample with resolution in the nanometre scale while acquiring creep curves, is exploited.

6.1 Hands-on Example 21: Creep Compliance Measurement of Viscoelastic Polymers

This hands-on example reports a work of Moeller [1].

Sample Preparation and Instrumentation The polymers studied, with elastic moduli between 20 MPa and 2.2 GPa, are Pebax® 3533, Kynar® 740, Kynar® 2800, and Kynar® 2750 resins, purchased from Arkema (King of Prussia, Pennsylvania). Pebax® 3533 is a thermoplastic elastomer made of flexible polyether and rigid polyamide; Kynar polymers are polyvinylidene fluorides (PVDF). The polymer

B. Cappella, *Mechanical Properties of Polymers Measured through AFM Force-Distance Curves*, Springer Laboratory, DOI 10.1007/978-3-319-29459-9_6

samples were embedded in epoxy cement, trimmed and cryomicrotomed in order to ensure a flat surface.

Measurements were performed with a MFP3D microscope (Asylum Research, Santa Barbara, CA) equipped with RTESP-type AFM probes (Veeco, Camarillo, CA) with radii between 10 and 50 nm, as well as four different types of LRCH probes (Team Nanotec, Villingen-Schwenningen, Germany) for measurements with much larger radii (530–761 nm). The tip shape was determined through scanning electron microscopy (SEM) images. The spring constant of the cantilevers was measured with the thermal noise method [2].

Creep measurements were performed using customized software, controlling the piezo elongation and the cantilever deflection via feedback loops. The protocol of the creep measurements consisted of a unique step load. The duration of the step load was varied between 200 ms and 90 s. The loading time was 20 ms, and the unloading time was varied between 20 ms and 60 s. Loads between 200 nN and 120 μN were applied. The analysis was performed by averaging several force–distance curves or creep curves acquired at different areas on the sample surfaces.

For comparison with AFM measurements, the elastic moduli of the polymers were determined using dynamic mechanical analysis (DMA). Measurements were performed with a RDA III instrument (Rheometric Scientific, Munich, Germany). Additionally, the modulus of Pebax® 3533 polymer was measured with instrumented indentation using a Hysitron TI 900 TriboIndenter (Bruker, Millerica, MA).

Creep Compliance Measurement of Viscoelastic Polymers The moduli of the studied polymers were determined at first through force–distance curves analysed with Oliver and Pharr method. For Pebax®, three sets of 5–10 force–distance curves were acquired on different regions of the sample with three different RTESP probes. The spring constants of the cantilevers were between 73 and 135.5 N/m. Care was taken to minimize uncertainties due to tilting of the cantilever, measured value of the cantilever spring constant and surface roughness. The area functions of the probes were calculated from SEM images acquired immediately after the indentations.

The measured elastic modulus was found to be between 16 and 44 MPa for the three probes, with standard deviations of ca. 10 % for measurements in one area and ca. 20 % for measurements in different regions. The elastic modulus determined with instrumented nanoindentation was 20.1 ± 2.4 MPa and the DMA value was 24.3 ± 1.2 MPa.

Apart the fact that Oliver and Pharr method is not suitable for the analysis of curves on viscoelastic materials, there are two main factors limiting the accuracy of AFM measurements: the size and the shape of the AFM tip. First, in order to minimize errors due to the tilt and to the deflection of the cantilever, a spherical tip should be used. Second, a larger tip permits to determine more accurately the area function and hence the modulus.

Measurements were repeated with three LRCH40 cantilevers, having a conical tip with hemispherical apex of radius between 530 and 761 nm. The cantilevers had spring constants between 58.8 and 67.4 N/m.

The second set of measurements yielded for the modulus of Pebax® 24.6 ± 2.9, 26.7 ± 2.7 and 30.3 ± 3.5 MPa. Hence, the use of larger radii enhances the accuracy of the measurements.

The author tested also the influence of indentation depth on the measured values of the modulus. The average value of the elastic modulus resulted not to change considerably when the indentation depth was reduced from 65 nm ($E = 24.4 \pm 1.8$ MPa) to 20 nm ($E = 22.8 \pm 5.7$ MPa), but the standard deviation increased significantly. The increase of the standard deviation is attributed to surface roughness effect: when the indentation depth is comparable with the surface roughness, the contact area is considerably affected by asperities.

The moduli of the three polyvinylidene fluorides (Kynar® 740, Kynar® 2800 and Kynar® 2750) were measured with four different LRCH cantilevers with very high elastic constants. Table 6.1 shows the results of the measurements. The curves were analysed again with Oliver and Pharr method.

It is evident that only the values for Kynar® 740 obtained with probes 2 and 3 are in quite good agreement with DMA values, whereas all other values are considerably higher than DMA results.

Since Oliver and Pharr method does not yield accurate results for viscoelastic materials, AFM creep tests were performed with the four studied polymer samples using the LRCH cantilevers. All step load measurements consist of a single step with a 20 ms ramp at the beginning.

Figure 6.1 shows the piezo elongation as a function of time on a Kynar® 2800 sample after a step load of 40 μN. Since the force and hence the cantilever deflection is kept constant, the piezo elongation is equivalent to deformation or indentation. Measurements of the cantilever deflection as a function of time showed that the cantilever deflection reaches a constant value after ca. 10 ms and that the deflection variation is smaller than 0.5 nm.

In order to calculate the elastic modulus from the creep curves, a suitable rheological model must be chosen. Three different models have been used by the

Table 6.1 Values of elastic moduli of the three polyvinylidene fluorides (Kynar® 740, Kynar® 2800 and Kynar® 2750). The values yielded by AFM measurements are compared with values obtained through DMA

	Elastic modulus (GPa)		
	Kynar® 740	Kynar® 2800	Kynar® 2750
Probe #1	–	1.63 ± 0.23	1.06 ± 0.15
Probe #2	2.59 ± 0.35	2.19 ± 0.19	0.97 ± 0.16
Probe #3	2.89 ± 0.42	1.96 ± 0.30	1.16 ± 0.16
Probe #4	3.30 ± 0.44	–	–
DMA	2.31 ± 0.16	0.87 ± 0.04	0.59 ± 0.03

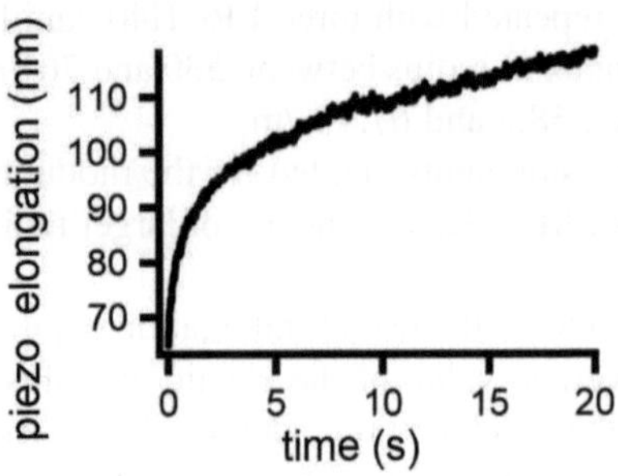

Fig. 6.1 Piezo elongation as a function of time on Kynar® 2800 after a step load of 40 μN. Adapted with permission from [1]. Copyright 2009. Wiley Periodicals

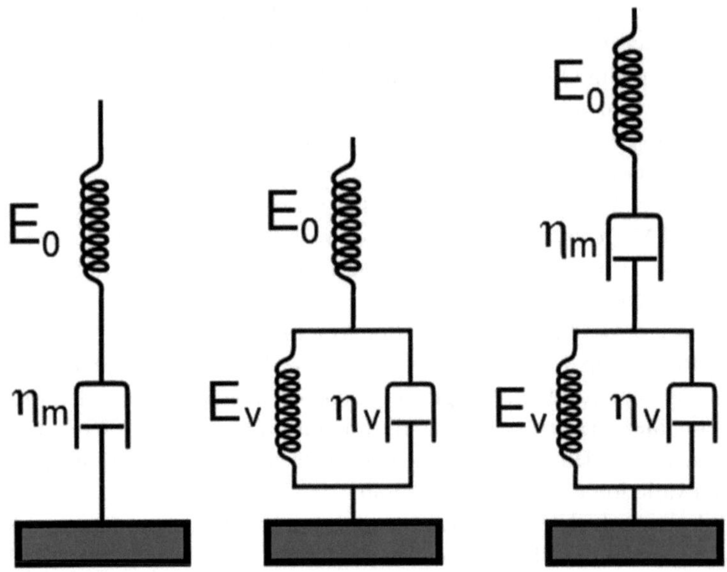

Fig. 6.2 Maxwell model (*left*), SLS model (*middle*) and the combined Maxwell–Kelvin–Voigt model used for the analysis of creep curves

author: Maxwell model, SLS model and the combined Maxwell–Kelvin–Voigt model, shown in Fig. 6.2.

The indentation depth or deformation D for a spherical indenter as a function of the moduli E_0 and E_v, the viscosities η_m and η_v, Poisson's ratio v and the tip radius R are given for the three models by [3]:

$$D^{3/2} = \frac{3}{4\sqrt{R}}\left\{\frac{4(1-\nu^2)}{E_0} + \frac{t}{\eta_m} + \frac{(1-2\nu)^2}{E_0}\left[1-\exp\left(-\frac{E_0 t}{3\eta_m}\right)\right]\right\}, \tag{6.1}$$

$$\begin{aligned} D^{3/2} = & \frac{3(1-\nu^2)}{\sqrt{R}E_0} + \frac{3}{4\sqrt{R}E_v}\left[1-\exp\left(-\frac{E_v t}{\eta_v}\right)\right] \\ & + \frac{3(1-2\nu)^2}{4\sqrt{R}(E_0+3E_v)}\left[1-\exp\left(-\frac{(E_0+3E_v)t}{3\eta_v}\right)\right] \end{aligned} \tag{6.2}$$

and

$$D^{3/2} = \frac{3}{4\sqrt{R}}\left\{\frac{4(1-\nu^2)}{\sqrt{R}E_0} + \frac{t}{\eta_m} + \frac{1}{E_v}\left[1-\exp\left(-\frac{E_v t}{\eta_v}\right)\right] + \frac{(1-2\nu)^2}{E_0}\frac{T_2 T_3}{T_2 - T_3}Z\right\} \tag{6.3}$$

with

$$Z = \left(\frac{1}{T_2} - \frac{1}{T_3}\right)\exp\left(-\frac{t}{T_3}\right) + \left(\frac{1}{T_1} - \frac{1}{T_2}\right)\exp\left(-\frac{t}{T_1}\right), \tag{6.4}$$

$$T_2 = \frac{\eta_v}{E_v}, \tag{6.5}$$

and

$$T_{1/3} = \frac{6\eta_m\eta_v}{E_0(\eta_m+\eta_v) + 3\eta_m E_v \pm \sqrt{E_0^2(\eta_m+\eta_v)^2 + 6\eta_m E_0 E_v(\eta_m - \eta_v) + 9\eta_m^2 E_v^2}}. \tag{6.6}$$

A creep curve on Kynar® 2800 (actually $4D^{2/3}/P_{max}$) is shown in Fig. 6.3 together with the fits: the continuous line represents the Maxwell model fit (Eq. 6.1), the dashed line the SLS model fit (Eq. 6.2) and the dotted line the combined Maxwell–Kelvin–Voigt model fit (Eq. 6.3). The best fit is obtained with the combined Maxwell–Kelvin–Voigt model fit, despite evident discrepancies for the initial 2 s, shown in the inset.

The combined Maxwell–Kelvin–Voigt model was used for the analysis of creep curves on all three Kynar® materials. Measurements were performed varying the step duration, which should not influence the measured value of the modulus.

Figure 6.4 shows the results, which should be compared with the already mentioned values yielded by DMA (2.31 ± 0.16 for Kynar® 740, 0.87 ± 0.04 for Kynar® 2800 and 0.59 ± 0.03 for Kynar® 2750).

For all three polymers, not only the modulus depends on the duration of the step load and decreases with increasing step duration but also the time constants are a

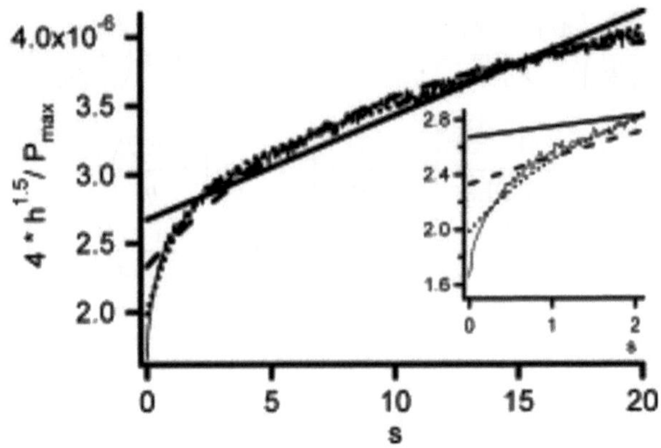

Fig. 6.3 Creep curve on a Kynar® 2800 sample. A step load of 40 μN with duration of 20 s was applied. Maxwell model fit (*continuous line*), SLS model fit (*dashed line*) and five-element model (*dotted line*) are shown. Reprinted with permission from [1]. Copyright 2009. Wiley Periodicals

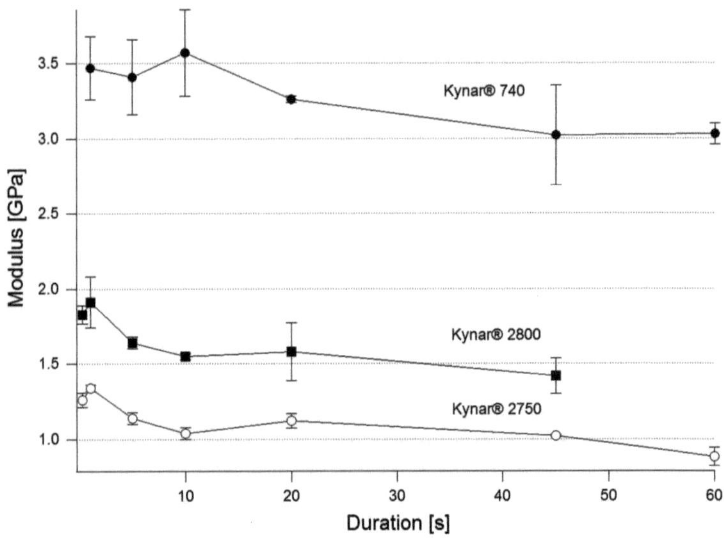

Fig. 6.4 Elastic moduli of the three different Kynar® samples as a function of the step load duration

function of the step duration. Also, moduli obtained from the creep curves are 30–50 % larger than the DMA values. Hence, the model is not correct.

The changes in moduli and time constants are due to the fact that the beginning of the creep curve, roughly the first 2 s, cannot be fitted properly. The author mentions two possible reasons for this discrepancy and disregards the fact that the step load is not a Heaviside function, assuming a 20 ms wide ramp as a good approximation of a perfect step. This is not in agreement with the results of Lu et al. [4].

The two phenomena considered by the author are (1) a more complex linear elastic behaviour, which needs a more complex model with more time constants, and (2) significant plastic deformation.

The deformation at zero unloading force or remaining deformation D_{rem} in a measurement with a long duration step is a good approximation for the total plastic deformation. By measuring D_{rem} with step loads of different duration and at different forces, the author found that:

1. The creep has a significant plastic contribution for all three polymers.
2. Reducing the load by a factor of two does not reduce plastic contribution significantly.
3. Reducing further the loads leads to less reproducible results and to overestimation of the elastic modulus.

As explained in Sect. 1.8.4, the contribution of plastic deformations to the creep can be eliminated with an experimental protocol consisting of two successive steps. Yet, such measurement was not performed in the work of Moeller.

6.2 Hands-on Example 22: Creep Compliance Mapping

In this second hands-on example, creep compliance measurements on homogeneous samples have been performed. Furthermore, a method for mapping the lateral distribution of viscoelastic properties of heterogeneous samples using AFM creep measurements is presented [5].

Sample Preparation and Instrumentation Three different samples were investigated: (1) natural rubber, (2) syndiotactic polypropylene (PP) and (3) a commercial epoxy adhesive.

The vulcanized natural rubber sample (McMaster Carr, Chicago, IL) had a Young's modulus of 2.3 ± 0.4 MPa.

The syndiotactic polypropylene was purchased from Sigma–Aldrich (product number 452157, St. Louis, USA). The elastic modulus provided by the manufacturer is between 211 and 244 MPa.

The epoxy adhesive was a two-component methyl methacrylate-based adhesive (Toolcraft Multi Power, distributed by Conrad Electronic SE, Hirschau, Germany). Resin and hardener of the fast curing adhesive were thoroughly mixed with a 1:1 ratio. The mixture was then placed on a microscope slide and examined after a curing time of 24 h. The elastic modulus was estimated at ca. 500 MPa.

Both the natural rubber and the syndiotactic polypropylene sample were also cryomicrotomed to ensure a smooth and flat surface.

All measurements were performed with a MFP3D BIO AFM (Asylum Research, Santa Barbara, CA, USA). Experiments were performed with the double step loading described in Sect. 1.8.4. The duration of the steps was always 150 ms.

Creep compliance measurements and data analysis were performed by using customized software, which is necessary for the double step loading.

The epoxy adhesive and the syndiotactic polypropylene were investigated using PPP-NCHR cantilevers (NanoWorld, Neuchâtel, Switzerland, spring constant $k_c \approx 30$ N/m). The natural rubber was investigated with an AC240TS cantilever (Olympus, Tokyo, Japan, $k_c \approx 1$ N/m). Both cantilever types had an approximately conical tip shape with a half-cone angle of ca. 20° and an apex radius of ca. 10 nm. The spring constants were determined by the thermal noise method [2].

Creep Compliance Measurements on Homogeneous Samples Creep compliance measurements were performed on homogeneous natural rubber and syndiotactic polypropylene, exhibiting very different viscoelastic properties. Natural rubber has a glass transition temperature T_g of ca. −70 °C, whereas T_g of PP is ca. 132 °C. As a consequence, the instantaneous modulus E_0 of natural rubber is two orders of magnitude smaller than that of PP. Also the retardation time τ is much smaller for natural rubber.

Creep curves were measured with a set point $F_0 = 50$ nN. The force applied during preload was $F_1 = 500$ nN for natural rubber and $F_1 = 650$ nN for PP; the magnitude of the main step was $F_2 = 200$ nN for natural rubber and $F_2 = 250$ nN for PP. The results are the average of 500 measurements per sample.

Creep curves are fitted with the equation for a conical indenter:

$$D = \frac{3}{8R}\int_0^t J(t-t')\frac{\mathrm{d}F(t')}{\mathrm{d}t'}\mathrm{d}t'. \tag{6.7}$$

The compliance J is given by the creep function $J(t) = J_\infty - (J_\infty - J_0)e^{-t/\tau}$ (Eq. 1.75).

In order to take account of the ramp in the step load, the force is written as

$$F(t) = (F_2 - F_0)\left(1 - e^{-t/T}\right), \tag{6.8}$$

where T is the rise time of the force step. Inserting the expressions for the force and the compliance in Eq. (6.7), the deformation due to the main step is obtained:

$$D = \frac{3}{8R}(F_2 - F_0)\left\{\left(\frac{1}{E_0} + \frac{1}{E_\infty}\right)\left(1 - e^{-t/T}\right) + \frac{e^{-t/\tau}}{E_2(T/\tau - 1)}\left[1 - e^{-t\left(\frac{1}{\tau} - \frac{1}{T}\right)}\right]\right\}. \tag{6.9}$$

Fitting creep curves acquired with Eq. (6.9), the viscoelastic parameters are obtained for both polymers.

For the natural rubber is $E_0 = 2.81 \pm 0.56$ MPa, which is in good agreement with the manufacturer value of 2.3 ± 0.4 MPa. The equilibrium modulus is $E_\infty = 1.88 \pm 0.23$ MPa, the relaxation time $\tau = 9.4 \pm 0.5$ ms.

For PP is $E_0 = 322 \pm 36$ MPa, $E_\infty = 122 \pm 15$ MPa and $\tau = 33 \pm 2$ ms. The manufacturer value for the modulus of PP is 211–244 MPa.

Discrepancies are attributed by the authors to plastic deformations. Indeed, measuring the normalized remaining indentation $D_{rem,2}/D_{max,2}$ for the two polymers, it is clear that the contribution of plastic deformations is larger in case of PP (67 % compared to 24 % for the natural rubber).

Creep Compliance Mapping Since the epoxy adhesive is not homogeneous in the micrometre scale, creep compliance measurements on this sample give information about the spatial variation of the viscoelastic parameters.

Figure 6.5 shows four maps obtained from the creep compliance measurements: the instantaneous modulus E_0 (a), the equilibrium modulus E_∞ (b), the retardation time τ (c) and the normalized remaining indentation of the preload step $D_{rem,1}/D_{max,1}$ (d). In all maps, there is a pronounced contrast between the

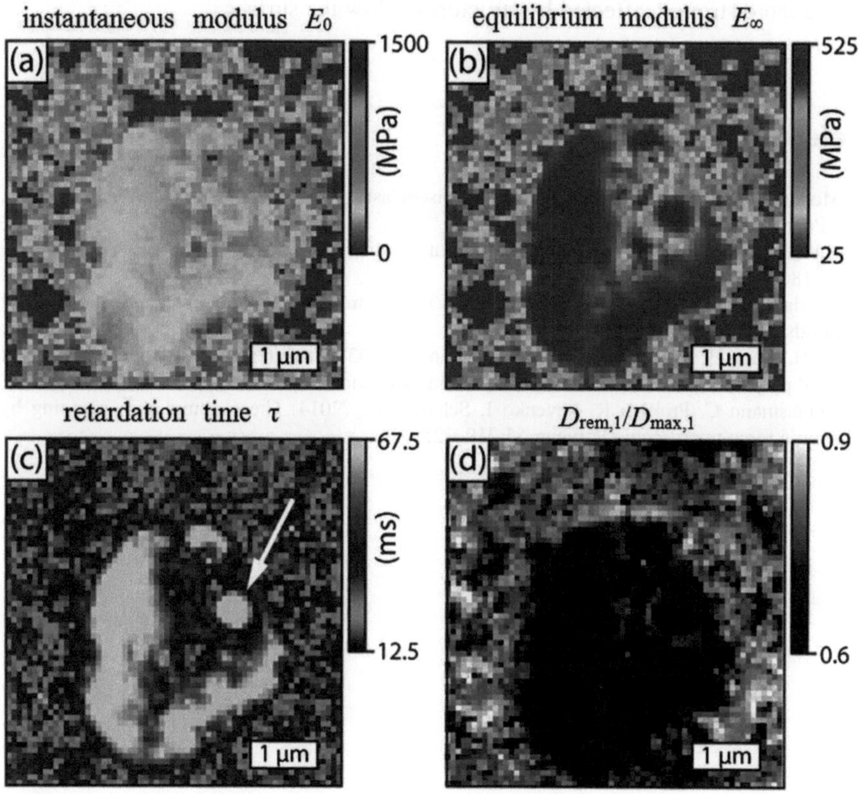

Fig. 6.5 Four maps obtained from the creep compliance mapping on the epoxy adhesive: (**a**) instantaneous modulus E_0, (**b**) equilibrium modulus E_∞, (**c**) retardation time τ, (**d**) normalized remaining indentation of the preload step $D_{rem,1}/D_{max,1}$. Adapted with permission from [5]. Copyright 2014. Elsevier

large central depression and the surrounding area. Creep curves were obtained with $F_0 = 100$ nN, $F_1 = 1600$ nN and $F_2 = 600$ nN.

The instantaneous modulus is more than two times larger outside the depression than inside (653 MPa and 316 MPa, respectively). The macroscopic elastic modulus estimated from manufacturer data is about 500 MPa, which is in the range of the measured values. The equilibrium modulus is even about eight times larger outside this depression than inside (309 MPa and 40 MPa, respectively). Also, the retardation time is considerably higher inside the depression than outside. The normalized remaining indentation of the preload step $D_{rem,1}/D_{max,1}$ is larger than 0.6 in the whole scanned area and is even larger (up to 0.9) outside the depression.

The creep compliance images show some submicrometre structures, for example, the one indicated by a white arrow in Fig. 6.5c, which cannot be seen neither in tapping mode images nor in usual force curve maps. This is due to the considerably larger indentation depth achieved in creep compliance measurements (around 300 nm compared with few nanometres in tapping mode and ca. 100 nm with conventional force–distance curves). The contrast in creep compliance maps is hence more strongly affected by structures below the surface.

References

1. Moeller G (2009) AFM nanoindentation of viscoelastic materials with large end-radius probes. J Polym Sci B Polym Phys 47:1573–1587
2. Hutter JL, Bechhoefer J (1993) Calibration of atomic-force microscope tips. Rev Sci Instrum 64:1868–1878
3. Vandamme M, Franz-Josef Ulm F-J (2006) Viscoelastic solutions for conical indentation. Int J Solids Struct 43:3142–3165
4. Lu H, Wang B, Ma J, Huang G, Viswanathan H (2003) Measurement of creep compliance of solid polymers by nanoindentation. Mech Time-Depend Mater 7:189–207
5. Braunsmann C, Proksch R, Revenko I, Schäffer TE (2014) Creep compliance mapping by atomic force microscopy. Polymer 55:219–225

Index

B. Cappella, *Mechanical Properties of Polymers Measured through AFM Force-Distance Curves*, Springer Laboratory, DOI 10.1007/978-3-319-29459-9

MIX
Papier aus verantwortungsvollen Quellen
Paper from responsible sources
FSC® C105338

If you have any concerns about our products,
you can contact us on
ProductSafety@springernature.com

In case Publisher is established outside the EU,
the EU authorized representative is:
Springer Nature Customer Service Center GmbH
Europaplatz 3, 69115 Heidelberg, Germany

Printed by Libri Plureos GmbH
in Hamburg, Germany